Secret US Proposals of the Cold War

Secret US Proposals of the Cold War
Radical Concepts in Military Aircraft

Jim Keeshen

Photos by Allen Hess

Crécy
www.crecy.co.uk

Crécy Publishing Ltd

Published in 2013 by Crécy Publishing Limited

Copyright 2013 Jim Keeshen

All rights reserved. No part of this book may be reproduced or transmitted in any form or by any means electronic or mechanical, including photocopying, recording or by any information storage without permission from the Publisher in writing. All enquiries should be directed to the Publisher.

A CIP record for this book is available from the British Library

ISBN 9 780859 791618

Printed and bound in China

Crécy Publishing Limited
1a Ringway Trading Estate, Shadowmoss Road, Manchester M22 5LH

www.crecy.co.uk

Front Cover Main Picture: *Convair illustration of their proposed nuclear powered bomber, the NX-2. The conventionally fueled J57s turbojets under wing indicate that this version of the NX-2 flew on the General Electric Direct Air Cycle nuclear engines. The J57s would be used for take-off and landings and to fly out of populated areas to cut down on the radioactive contamination that would come from the nuclear engines once they were turned on.* (Scott Lowther)

Front Cover Inset: *In-house Douglas Model 1211 Parasite Bomber used in the B-52 competition carrying its mission specific pods that could hold fuel, equipment or bombs. The proposed bomber could also hold a variety of parasites under wing, including the Douglas XA4D, the Convair XF-92A along with a variant of the Northrop Snark.* (John Aldaz collection)

Back Cover Main Picture: *Illustration of the Douglas XB-42 bomber from the proposal brochure of 1942. This propeller powered medium bomber developed during the war was caught in the technological shift to jet power. After its inline engines were replaced with turbojets, this aircraft became the first American bomber.* (National Archives via Ryan Crierie)

Back Cover Inset Top: *Nemoto solid wood display model of the B-45C Tornado. This aircraft was designed from the beginning to be a jet powered bomber. Notice the WWII influence of the glass nose, straight wings and tail gunner position. Unlike the XB-42, the Tornado went into production, becoming the first American jet bomber to be in service with the Air Force.* (Author)

Back Cover Inset Center: *Illustration from the Curtiss-Wright XP-87 Blackhawk proposal brochure of the all-weather night fighter. This cutaway view of the Blackhawk shows the placement of the fuel tanks, engine and landing gear. Failure to receive a contract for this aircraft brought down the Curtiss-Wright Company.* (National Archives)

Back Cover Inset Bottom: *North American NA-239 with its impressive floating wingtip fuel tanks for extended missions. This was an attempt to develop a bomber capable of flying to the USSR and back. The concept never succeeded but it did evolve into the XB-70.* (Author)

Front Flap Top: *Illustration from the Boeing MX-1712 brochure demonstrating the flexibility of their proposed supersonic medium bomber from 1951. Shown here are three MX-1712 fuselage cutaways indicating three different configurations; as a bomber, a missile carrier and a photo-reconnaissance aircraft.* (National Archives)

Front Flap Bottom: *Model of an early flying wing, the N-1M "Jeep" by Jack Northrop. The "Jeep" had outer wing panels that could be repositioned to give different degrees of anhedral. It was used to help develop its larger brother, the XB-35.* (Barry Webb model)

Previous Page: *The Lockheed L-227-1 was a proposal for a new fighter design in 1952. Estimated performance included a combat ceiling of 52,600ft (16,032m), combat radius of 350nm (648km) and rate of climb 47,000ft/min (14,326m/min). This model was the precursor to the Lockheed F-104.* (Author)

Opposite: *Although in USAF markings, the Navy commissioned Republic to produce a turboprop fighter. With modifications to the fuselage, wing and tail, an F-84-35-RE became the XF-84H. When the stubby propellers reached supersonic speed they would let out a hideous high-frequency squeal. This former Thunderstreak was re-named "The Thunderscreech."* (Allen Hess model)

All photos are copyright Jim Keeshen and Allen Hess unless otherwise credited

Table of Contents

Acknowledgements 7

Introduction 9

Preface　　　Modelers and Model Making 13

Chapter 1　　Models and Artwork..18

Chapter 2　　The Bombers 36

Chapter 3　　The First Jets..86

Chapter 4　　Fly Navy!...126

Chapter 5　　Vertical Flight and Other Concepts.....................................148

Conclusion 167

Bibliography...169

Index 171

To a USAAF B-25 pilot who flew with the British in North Africa, and luckily married a woman who patiently put up with his airplane collection—to dad and mom

Acknowledgements

CREATING THIS KIND OF BOOK requires the help of many. On the publishing side, my deepest thanks go to Jeremy Pratt and Gill Richardson of Crécy Publishing in the UK and to Dave Arnold of Specialty Press in the US. Their support, enthusiasm and encouragements cleared my path to go on with the tasks at hand.

But the one who started it all was Mike Machat, former Acquisitions Editor for Specialty Press, who, during lunch one day, asked me a rhetorical question, "Who could write a book on concept models?" I named a number of collectors who could easily qualify, but then realized Mike had already made up his mind. For giving me the opportunity to write this book, I am most grateful.

The nuts and bolts part of this book, the researching, constructing information, locating and photographing models also involve many individuals. I have the good fortune to know a number of friends who happen to share the same interests in aviation. Here are most of them:

Two close friends happen to be aerospace engineers from Northrop and North American Rockwell, Ron Monroe and Allen Hess, respectively. These two people need special mention. Ron was an aircraft performance analyst in Northrop Grumman's Advanced Design Group. He was amazing in searching out materials from his library, and the Internet, but one of his biggest contributions was his ability to find, call, and recruit retired engineers who would answer questions on long forgotten projects that could not be found in any archive or web page.

Allen Hess designed wind tunnel models for North American Rockwell. He is also an international award-winning model builder who happens to be a professional photographer. When I asked him to shoot a couple of his models for the book, he was happy to oblige. To my great fortune – and to the visual quality of this book, he then offered to shoot any other models for the book. Allen went with me across town or across the nation visiting museums and collections, setting up his portable studio and shooting most of the pictures in this book. Add to this his wealth of knowledge about these models (not to mention doing expert repairs on a number of broken ones), and you begin to appreciate how blessed I was to have Allen's help.

Tony Buttler, author of numerous aviation books, became my literary mentor. He invited me to the National Archives in Washington, D.C., where he literally walked me through the process of mining for nuggets in thousands of boxes filled with aviation ephemera. Thank you, Tony, for your generous help and guidance, for answering my many inquiries and for making numerous suggestions to the text.

Sir George Cox, a man of many hats, fortunately for me, three of which were most helpful; his eclectic model aviation collection (more than likely the largest in Europe), the vast knowledge of its contents, and his editorial skills. Sir George generously opened his entire collection for this book, patiently supplied my multiple requests for specific pictures, and kindly took the time out of his very busy life to review and improve the manuscript. Thank you, George, for fitting me into your busy schedule to help.

Allyson Vought, the Grande Dame of secret projects and concept models in the United States, started her collection back in the 1960s, amassing one of the largest, if not the largest collection of concept models. Her vast cache inspired the rest of us collectors. Allyson's expertise and connections in the collecting world proved very helpful for this book.

While working at the National Archives I met Ryan Crierie. Ryan is an amazing researcher with an uncanny knack for finding some of the most unusual material for this book. Ryan also scanned and cleaned a number of the images to be used in this book.

While researching in Washington D.C., I met Caroline Sheen, Photography and Art Editor for the *Smithsonian Air and Space* magazine. She in turn, opened the doors at the National Air and Space Museum (NASM) and introduced me to Christopher T. Moore, Museum Specialist. Chris gave special access to view and take pictures of the extensive model collection at NASM. And to my great delight he then took us to the Silver Hill Restoration Facility to view the stored historical aircraft waiting their turn to be exhibited. Thank you again, Caroline and Chris.

At the San Diego Air and Space Museum, Tony Beres led me to Terry Brennan, Curator and Director of Restoration. We were well taken care of by Al Valdes, Assistant Curator for Collections. Special thanks to Al for giving us access to the Ed Heinemann Collection which SDASM holds and for making the extra effort to find the lost and unknown seaplane wind tunnel model in the lower catacombs.

In Long Island, New York: Larry Feliu, Director of the Northrop Grumman History Center, was most generous with his time and his facility. He gave us access to anything we needed, including the many unknown Grumman models that hang overhead in his office. His History Center is a wealth of information that seems to be underappreciated. Larry introduced me to Lynn McDonald who connected me with his group, the Long

Island-Republic Airport Historical Society. Lynn was instrumental in getting the society's president, Leroy E. Douglas, and treasurer, Charlie Bowman, to come down to the Republic Airport Terminal to open the display cases containing some exceedingly rare historical models. Lynn, Leroy, and Charlie were very generous with their time, spending the entire day opening cases, cleaning off the dust and setting up the models to shoot.

At the former grounds of Mitchel Air Force Base in Garden City, New York, Joshua Stoff, Curator of the Cradle of Aviation Museum took time out of his very busy schedule and personally chose models from his impressive collection for us to photograph. Joshua also gave us complete access to the rare Republic archives.

At the Republic Airport, in the remnants of the original Republic hangars, the curator of the American Airpower Museum, Larry Starr, gave us a wonderful tour of his fine museum and, of course, allowed us to shoot his collection of unique Republic concept models.

Besides opening up their museum vaults, Al Valdez, Larry Feliu, Joshua Stoff and Larry Starr, also gave me an appreciation of how personally dedicated one has to be to run a museum, find financing, set up exhibits and just keep the doors open for the love of aviation history. These are passionate and dedicated individuals.

Also in Long Island I was introduced to Bill DiNoia and Brian Aubin, former heads of the Grumman model shop and now owners of Creative Models. They were more than generous with their stories, time, photos, and technical information on the history of the Grumman model shop. Special thanks for your kind attention. They introduced me to one of the leading collectors on the East Coast, Larry McLaughlin.

Larry McLaughlin pulled out all the stops. He invited me to his home, allowed me to pick and choose any rare airplane I pleased from his fabulous collection. Again, the generosity and hospitality I kept encountering from the collectors was simply overwhelming. Larry went the extra mile and pulled models out of storage units before my visit, just so I could have more to pick from. He even provided me with rare model photos. Thank you, Larry, for all your kind help and attention.

On the west coast, Greg Barbiera did exactly the same thing when I called and asked to photograph his collection. Greg has some of the rarest models out of North American and Northrop. Thanks, Greg, I truly envy quite a number of your gems.

Also on the west coast, Jonathan Rigutto was very helpful and generous with his collection. I have known Jonathan for some time and truly appreciate his kindness. He not only trusted me with his rarest models, but he also lent me hard to find books and papers on cold war concept models out of his library.

John Aldaz has a wonderful collection and allowed me access to it. What is impressive about John is his knowledge about concept models and more than most, could provided detailed information on any of his – or anyone else's – aircraft model. He most certainly kept his promise to help me make this book the best possible.

Thank you to Douglas Frie and Steve Robles, engineers at Republic and Northrop Grumman, respectively, who provided information on the early concepts; to Mike Kellogg for the use of his Cruver ID models; and to Mike Peters who helped me put together a number of the graphics in Photoshop.

A very special thanks to Tony Chong, head of the Northrop Grumman Display Model Shop and my content editor, who ensured technical accuracy and helped keep the book on track right from the start. In addition, Tony generously contributed his Northrop artwork and photos to this project.

And the last word of thanks goes to Una Seo. Without her support, insights, advice, prodding and organizational skills, this book would never have made it. She is my biggest fan, and I am hers. Her warm encouragement kept me going.

There are many more people who helped in many ways with this book so please forgive me for not mentioning you. Nonetheless, you are all very much appreciated.

Introduction

THE ADVANCE OF AVIATION has produced many fantastic designs—not all of which have seen the light of day. For every military airplane that went on to enjoy a successful career with the armed forces, several others never progressed beyond the prototype stage and scores more that never left the drawing board. Today our knowledge of these projects is limited to surviving company documents, artwork, and models. Collectively, this information provides a fascinating picture of man's quest to go higher, faster, and farther.

This book concentrates on the period between 1945 and 1965: two decades that saw the greatest proliferation of new ideas that changed the face of military aviation. During this era, new airplane designs were created continuously by the West in an endeavor to counter the mounting Soviet threat of the Cold War. These new aircraft were designed for varied roles. Some were simply intended for research, pushing the boundaries of knowledge and aiming at technological goals. Others were conceived to support ground forces, patrol the oceans, defend U.S. airspace, or penetrate Soviet defenses.

The book examines some of the period's more unusual concepts as presented in models and drawings created by the model shops and art departments of U.S. airplane manufacturers. It also takes into account some conventional airplanes to set the more exotic designs in their historical context. Included with in-house proposal models of aircraft that were never built are unusual prototypes, display models, ID models, and even some plastic models from the hobby industry to help fill out the story of this amazing time period of aircraft development.

Douglas Aircraft Corporation built this large in-house wood and metal model of their Model 1240 parasite bomber to compete as the replacement for the B-36. Originally dubbed the "All Purpose Bomber" this aircraft could be configured to carry a number of different fighters, missiles, bombs and pods. This particular example shows an observation pod hung in the center section. After a reconnaissance mission the photographs taken could be developed in a fully equipped dark room and analyzed during the homeward bound flight (Author)

The Design Process

To better understand how and where these models and drawings came about, a quick overview of how aircraft are conceived and built is necessary. In the United States, aircraft designs emerge from one of two different trains of development. One is by request of the government for design proposals to meet a particular need. The other comes from the manufacturer to the government, either as proposed improvements to an existing design or as suggestions for an entirely new capability from which the armed forces might benefit.

Most aircraft designs do not follow an easy road to completion. Many twists and turns along the route affect their line of development or even their demise, some technological, some financial, some political, and some inexplicable. Once ready, presentation of these new designs is made to the government as a compilation of cost analysis, IOC (initial operational capability), estimated performance data, along with graphics and sometimes three-dimensional models, in the hope of obtaining a contract.

An example of how tortuous this process can be is the Northrop P-61 Black Widow of World War II. It came as a request from the government when U.S. Army officers witnessing the Battle of Britain, realized that interception of enemy aircraft would be quicker if radar detection was done in the air instead of vectoring fighters from the ground. Shortly thereafter, in the fall of 1940 a specification for a new fighter was issued to the industry. One of the requirements was to carry the then-new radar used for interception. The P-61 preliminary design was submitted on 7 December 1940, and Northrop received the contract for production on 30 January 1941.

It was not an easy task. Unlike conventional fighters, the new plane had to carry a radar-operator as well as a pilot. It also needed a radar scanner with an unobstructed forward view. Before the P-61 was built, several design iterations were made by Northrop. Each time the weight went up, so more power was needed, so a bigger engine was installed, so the weight went up, and so on. This is a typical design process; the engineers make design compromises and find new design breakthroughs, all the while keeping in mind that the airplane cannot be overweight, underpowered, or overly expensive. And along the way, each iteration may be rejected within a couple of days or a couple of weeks. Meantime, the process is often complicated by changes in the requirements as military thinking shifts. Eventually the designers come up with an airplane that meets all the specifications and requirements.

The length of this process is illustrated by the fact that the P-61 Black Widow was eventually delivered in 1944, just in time to see action in the very last stages of the war.

There were other designs that took even longer, started at the beginning of the war and never completed until after the end of the conflict. The Consolidated Vultee B-36 and the Northrop B-35 were both ordered in 1941 but did not take to the skies until 1946.

This design process has actually gotten longer over the succeeding decades. For example had the P-51 Mustang (one of the outstanding WWII fighters) taken as long to develop as today's more modern fighter (the F-35), it would not only have missed the War, it would have missed the Korean conflict too and only have been in full squadron service in time for Vietnam!

Once the aircraft is designed and under construction, the design engineers are already turning to the next project, either thinking and drawing up how the new aircraft can be further enhanced or what might succeed it.

A Word About Grumman Models

Grumman-produced models are among the most beautiful models ever made. The quality of the craftsmanship is superb. Grumman is the only company that made early display models in two ways: as a blank wood form and as a highly finished, painted model. Examining one of their wood blank models closely, it becomes apparent that great care was taken in constructing the various wood parts, shaping and blending them to make an exact representation of the proposed design, or even an existing aircraft. An example of this quality is shown on the right.

Northrop XB-35 Flying Wing, an in-house resin model of the Flying Wing powered by four Pratt & Whitney R-4360 turbo-supercharged radial engines, each driving a set of eight contra-rotating propellers. (Greg Barbiera Collection)

Introduction

Poplar or basswood was used to construct this in-house variant of the proposed Grumman Design 97 from the 1950s. The craftsmanship is superb on this 17 inch (43 cm) long model. The wood surfaces are varnished and polished and the canopy is a solid piece of shaped and polished Plexiglas. (Larry McLaughlin collection)

For comparison, here is a finished example of the in-house Grumman Design 97 proposal. Note the difference between the verticals in both planes. It makes an impressive presentation displayed with its highly polished, professional finish.
(Northrop Grumman History Center)

11

A Grumman wind tunnel model of the F9F Panther, called a "semi-span" is made from a combination of aluminum and stainless steel. The length of this model measures about 12in (30cm). (Northrop Grumman History Center)

How can new technological breakthroughs be exploited? What could be done in a few years time that couldn't be done today? What new "enemy" threats are emerging and what needs to be done to counter them? What matters more: speed, armament-carrying capacity, or range? Studies are made, think tanks are consulted, and discussions are held with military personnel. This, hopefully, gives the company a competitive advantage over the other aerospace competitors.

Once the manufacturer has what it feels is a potentially attractive aircraft, it meets again with government officials to sell its idea. It may be a radical departure from what was done before, or it may be a modification of an existing design. The manufacturer may write informational reports called white papers, describing the need for the new aircraft, and how it can be used. If it can generate interest, the manufacturer may receive a request for more information, or a request to make a presentation to other officials. Brochures with conceptual art and graphics may be printed. And when the manufacturer feels there is enough interest, it may even go so far as to build a presentation model, bringing the concept to life.

At this stage, the government may choose to open the competition to the industry to see what the other companies can propose for the same requirements. Other companies' designs may look very similar or very different, or the designs may go through iterations as the thinking progresses. This brings about many more different layouts and it was not uncommon for a few handmade models to land on an official's desk: an enticing representation of the aircraft displaying USAF or US Navy colors.

Models and graphics of these new designs were thus made to help the engineers visualize the aircraft to verify their ideas for construction, to sell the product, and to encourage the military to consider the possibilities of what might be. Although quantities of graphics and accompanying documents survive to this day, not all are readily accessible to the researcher.

Author Tony Buttler said it best in his book, American Secret Projects: Fighters and Interceptors 1945 to 1975: "Inevitably, much of the information—highly secret when it was originally produced and hence shut away from public scrutiny—has been lost over the years, but I am sure that much still remains to be discovered, buried in archives, personal collections, or simply people's memories." For this reason this book makes no pretence of being exhaustive. The models and artwork shown in this book are more of a glimpse of known artifacts, recent discoveries, and with some models, pieces of puzzles yet to be identified and correctly placed in their historical lineage.

Preface
Modelers and Model Making

PART OF THE RESEARCH for this book involved seeking information on certain models and interviewing model makers from the aerospace industry. A number of these interviews were so compelling that it was decided to include a brief look at professional modelers and the story behind their craft. Here now are two model makers from the old Grumman Corporation in Long Island.

Bill DiNoia and Brian Aubin started as apprentice modelers in the early 1980s and worked their way up to section heads of the Grumman Model Shop until the facility closed in 1994. They now own and run their own successful business, and are partners of Creative Models and Prototypes in Hicksville, Long Island. Their model shop is not far from their former employer, now called Northrop Grumman Aerospace Systems.

Bill and Brian are locals; Bill grew up in New York City and Brian in West Islip. Both came from homes that fostered the hobbies of building model trains and model airplanes. Both loved to work with their hands. Bill began as a jeweler, sculpting and casting rings and broaches. Brian studied and received his degree from State University of New York, Oswego in Industrial Arts Education. They eventually met at the Grumman Model Shop where they were each hired in 1982.

They joined Grumman at a time when the shop was at the height of its long production history and met the other craftsmen, patternmakers, machinists and woodworkers who, at that time, already had 30 to 40 years of experience.

When asked what it was like starting work at the shop, Bill smiled and quickly pronounced, "It was a weird place!" Brian wholeheartedly agreed, "The Grumman model shops had about 80 employees dedicated solely to designing and building models. These were serious craftsmen; many of them came over from Germany after World War II. They were hard people to get to know, they kept themselves to themselves."

The Grumman Model Shop consisted of three divisions: the machine shop, the metal model shop, and the wood model shop. The machine shop worked on the large metal models for wind tunnel testing and it also specialized in producing customized, one-off metal pieces for real aircraft and the ground support equipment for those aircraft.

The metal model shop made smaller metal models and specific metallic pieces used on wind tunnel models that needed precise contours.

The wood shop produced presentation display models, engineering design aides, and low-speed wind tunnel models. These wind tunnel models were usually constructed in combinations of wood and metal, and sometimes fiberglass. Besides aircraft, the Grumman wood shop also produced various other models such as space satellites, rockets, weapons, and anything else that Grumman needed to show in miniature form.

Referred to as the "Country Club" throughout the company, the three divisions of the model shop (machine shop, metal shop, and wood shop) were housed inside Plant 5 of the Grumman factory. These created the core of the model development, but throughout the plant there were upwards of 20 satellite shops, known as cells, where models were constructed under high-security conditions.

Upon starting work, the young modelers were given apprentice jobs—making stands and plaques for the models, before yet being allowed to make a single model. The breakthrough came when one of the German section supervisors discovered that Bill and Brian could sculpt shapes and problem solve a modeling project to completion; talents they had honed during their boyhood model-building experiences. These abilities to sculpt shapes and figure out how best to accomplish a three-dimensional project are invaluable skills for the model maker. When the lead supervisor found this out, both Bill and Brian were promoted to actually making models.

Bill and Brian began building A-6 metal wind tunnel models. Both went on to work on radar cross-sectional models (RCS) of designs from other aircraft companies, such as the Lockheed SR-71 and the F-117.

As their careers grew, they worked their way up to the top supervisory positions. Bill and Brian were made Lead Man of the shop and eventually, Section Lead, the highest position in the model shop.

Both of these talented craftsmen stressed that the one thing they truly enjoy about making models is the constant learning and problem solving that takes place in the model shop. This was true during the beginning of their careers, and to this very day they are learning new techniques and ways of making models within their own business. Both seem to truly enjoy the art form of creating beautiful models and working with their hands.

Model Manufacturing

Many types of models are used in the aerospace industry; there are models for display, for sales promotion, for research and study, such as a wind tunnel model, or engineering concept. All have slightly different ways of being made. In the past, if a "one-off" (a single model), or only a few examples of a design were to be made, they were usually hand carved out of wood, and given detailed painting and markings. In other cases, models produced in multiples, or mass quantities, a wooden, clay, or foam master is created. A mold is taken from the master using materials such as fiberglass, or RTV (room temperature vulcanized) rubber, and copies are produced using materials such as urethane plastic, fiberglass, resin, or even metal.

The wood used to make models can vary from bass, gelutong, pine, maple, mahogany, birch, poplar, and sometimes balsawood. These wood models and masters are made by a modeler or wood worker.

Low-speed wind tunnel models are fabricated from a combination of wood and metal. High-speed wind tunnel models are mainly, if not exclusively, all metal; aluminum and/or stainless steel. In pre-computer days, wood or plastic patterns were made and then copied with a milling machine into metal parts. These types of models require a machinist to produce the various parts.

The actual process of model making has gone through an evolution over the years. The first models were made primarily from wood, sometimes with metal, wire, Plexiglas, or plastic pieces added for detail. But the art of sculpting a model is now aided by computer technology. Once crafted by hand, models are now formed easily by computer-assisted tools. Some models today are still made by hand, but instead of wood, high-density foam modeling board is used.

The photographer for this book, Allen Hess, is himself a retired wind-tunnel-model designer who worked for North American Rockwell from the late 1960s to the late 1990s. He recalls the patternmakers who worked for the model shop being eventually replaced in the 1990s by numerically controlled machine tools and a device called a Stereo Lithography Apparatus (SLA). The SLA is a three-dimensional manufacturing method that creates models within a tank of liquid UV-curable resin. The computer-controlled ultraviolet laser shoots a beam into the curable resin. The resin reacts and hardens into very thin layers, and layer upon layer, the model starts to form, suspended in the liquid resin. The result is a blank model with multi-faceted sides, something akin to an airplane made out of Legos. Upon removal of the blank from the resin, it is cleaned, the facets sanded into smooth form, primed, and readied for paint. Eventually, computer technology advanced to the point where a model is made with perfectly smooth contours, and now there are several different processes for generating 3-D models by computer.

An example of a blank model of the North American Rockwell X-31 made from the Stereo Lithography Apparatus (SLA) process. Note the model is in two pieces, for the SLA was not big enough to accommodate the entire length of the fuselage. (Allen Hess collection)

The faceted sides of the SLA model have been sanded smooth and primed. The model is now ready for the final coloring and decals. (Allen Hess collection)

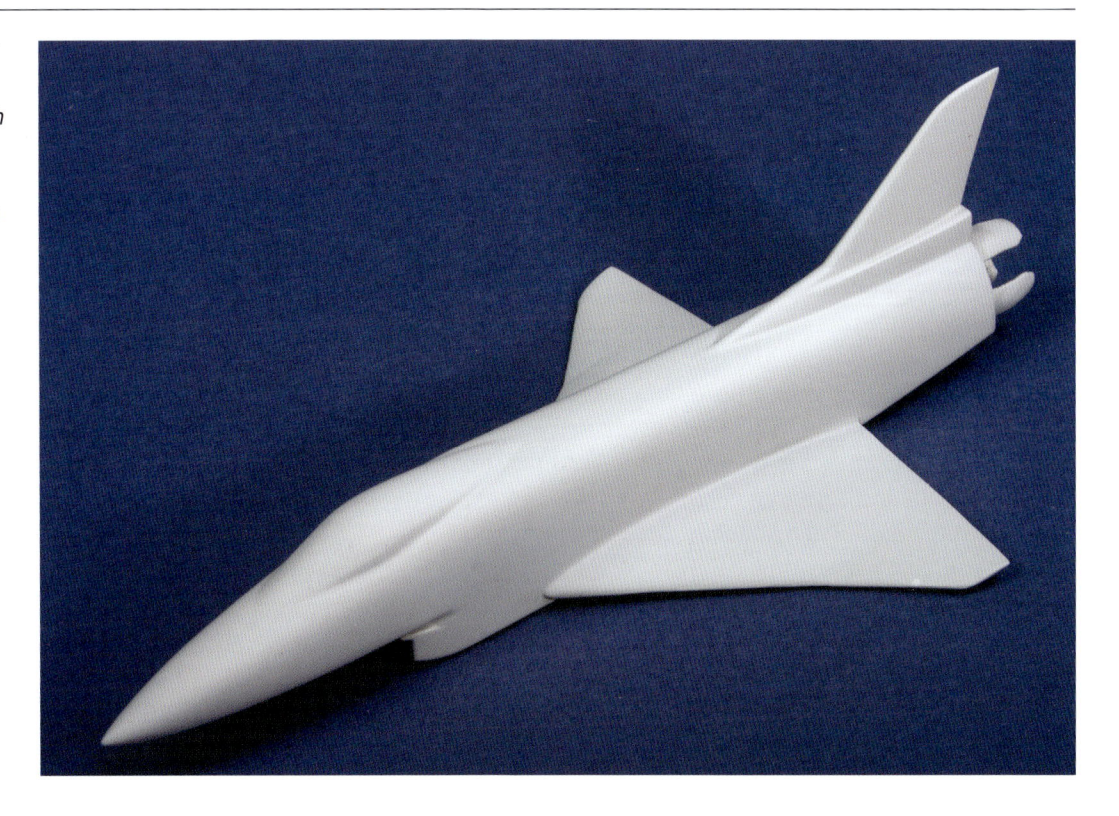

Below: *The final product, a finished SLA model of the North American Rockwell X-31, adorned in its final coat of color with decals added.* (Allen Hess collection)

A Final Word About Model Shops

As already mentioned, most U.S. aircraft manufacturers kept full-time, in-house model shops during this period. These shops typically employed dozens of skilled craftsmen, who were responsible for fabricating the elaborate wind tunnel models required by research and engineering departments. In addition, the shops produced conceptual, proposal, trade-show, and other display models for sales and marketing applications. These pieces often competed for the attention of a single customer—the Pentagon. Given the very lucrative nature of any potential aerospace contract, little consideration was given to their fabrication costs.

Prior to the digital age, a handcrafted model was the only tool capable of providing a three-dimensional, visual reference of any aircraft concept or design study. The resulting creations were nothing less than unique, unsigned works of art; highly accurate and finely finished.

Other types of replicas had to be produced inexpensively and in large quantities. These included ID recognition models and desktop promotional gifts, provided by airplane manufacturers to pilots and just about anyone involved with a particular type of aircraft. These typically mass-produced models were contracted to a variety of specialized companies, both in the United States and abroad. Best-known among them is Topping, whose durable, injection-molded creations could be found on just about every desk or shelf in the aerospace industry.

When North American Aviation became famous for its record-shattering X-15 rocket plane in the early 1960s, more than 15,000 Topping X-15 models were produced for everyone who participated in the program. By contrast, only three real X-15s were ever built and flown.

In addition to Topping, companies such as Precise, Pen-Wal, Blane, Southwick, Allyn and Hyatt shared in the good fortune of a vibrant U.S. aerospace industry. Topping was helped by generous deduction laws, which essentially allowed such promotional gifts to be funded by taxpayers. This loophole ended in 1964, at the insistence of Secretary of Defense Robert Strange McNamara, in his all-out assault on military procurement costs. Contractor gifts were hence limited to $5 in value, when the cost of a Topping model averaged about $9. Prior to that restriction, promotional models were given freely, or occasionally sold at cost through employee gift shops across the industry. Consequently, production numbers were impressive: Topping delivered more than 135,000 models of the C-135 jet transport, 200,000 Atlas ICBMs, and an unbelievable 250,000 F-104 Starfighters, among many others.

Foreign makers also produced models for U.S. aircraft manufacturers. Typically, these were high-quality, low-volume productions, made possible by substantially lower labor costs. Japanese labels, such as Nemoto or Sommer & Co. can be found on a variety of Grumman, Fairchild, and Martin models, all exquisitely finished. The Scandinavian metal foundry Fermo produced numerous aluminum models, and in the Netherlands master model-maker

Grumman OV-1 Mohawk in-house wood model (right) next to a plastic Topping Mohawk. The in-house Mohawk is slightly larger than the 1/48-scale Topping. The Topping has a clear canopy with two figures sitting inside while the in-house has its canopy painted to simulate glass. (Author)

Matthys Verkuyl made similar models for Lockheed, Republic, Northrop, Ryan, and others. In the UK, companies like Minavia, Woodason and Westway produced beautiful models for various aircraft manufacturers (and airlines), to be followed later by Space Models.

Few model makers survive today. Of the entire aerospace industry, only Northrop Grumman, Boeing, and Lockheed Martin still keep an active, in-house model shop. Their expert staff combine old-time craftsmanship with modern technology, such as the Stereo Lithography Apparatus (SLA). Independent model companies have also turned to the computer to stay competitive. Pacific Miniatures makes custom tooling and molds directly from the original aircraft manufacturer's CAD files, to ensure the highest level of accuracy. Artwork and decals are created using the same data files. Even packaging design is optimized by digital processes.

Model making has surely evolved from the days of hand carved wood forms. Wind tunnel models have been replaced by computational fluid dynamics. Proposals and presentations now use 3D graphic simulations to great advantage. The art of model making, for the most part, has come and gone within the twentieth century, leaving behind a legacy of truly spectacular creations.

This book is made to spotlight an often under-appreciated source of research and history, the model. The surviving model aircraft, preserved in museums or held by collectors, are testaments and sometimes puzzles for the research that is being made into fighters and bombers developed during the Cold War. Not enough has been said to stress how important and valuable they were at the time, and how even historically significant they are today. This book attempts to give the reader a better appreciation of some of the surviving models from the Cold War era, a time of angst, hope and tremendous aeronautical advancement.

Another experimental model made by Topping was of the North American X-15. These models were massed produced to be sold in the company store or given as gifts or commemoratives to the employees, military and pilots that were involved with each program. Over 15,000 models of the X-15 were made. (Allen Hess collection)

Topping Models not only made replicas of production aircraft, but it also produced experimental airplanes as well. The white VTOL aircraft with the silver rotors is the Curtiss-Wright X-19. The silver one with the shrouded rotors is the Bell Aerosystems X-22A Tri-Service Research V/STOL Aricraft. (Ron Monroe collection)

In-house model of the Douglas XB-42A demonstrates the first variant of the XB-42 which incorporated the small jet pods in order to improve performance. This model truly reflects what was happening to propeller driven aircraft at the dawning of the jet age. (John Adaz collection)

Chapter 1
Models and Artwork

THIS CHAPTER will look at how an aircraft manufacturer used models and artwork to sell an idea to the military. Also, to better understand how the U.S. aerospace industry developed during the Cold War, a look back at the 1940s is in order. We will explore how German advanced designs affected and shaped the emerging aerospace industry. But first, let us start with one specific example of a new bomber design proposed through models and artwork and how it got caught in the technological shift of the jet age.

Douglas Presents a New Advanced Bomber Design

Artwork and models played a key role for the Douglas Aircraft Company to land a contract for the XB-42 prototypes in 1943. Douglas wanted to sell this plane – incorporating many new features at the time – as a versatile long-range medium bomber. The proposal brochure illustrates this aircraft in its different roles as a surveillance plane, attack plane, or torpedo bomber. It even provides a visual comparison to some nebulous, current, heavy bomber to make its case of how more efficient it could be. The XB-42 showed promise, but it was a propeller aircraft caught at the beginning of the jet age. Upgrades were made to save it. The XB-42A was given better engines, Allison V-1710-133, and two Westinghouse 19B-2A turbojets were mounted under wing. The aircraft gained faster speed, but at the sacrifice of longer range; losing approximately 650 miles (1,046km) compared to the XB-42.

The Army Air Force cancelled the Mixmaster but was curious to see if the aerodynamically clean airframe could be converted into an all jet version. Two XB-43 Jetmaster prototypes were ordered, powered by General Electric TG-180 axial-flow turbojets, later re-designated J35-GE-3s. The XB-43 first flew on 17 May 1946 becoming the first American all jet bomber. But its performance was not enough, and was passed over in favor of the North

Illustration from the Douglas XB-42 bomber proposal brochure of 1943. The artist is Lieutenant Commander Arthur C. Beaumont who later taught or influenced the next generation of illustrators who created the artwork for the box tops of the plastic model hobby industry like Revell and Aurora. (National Archives via Ryan Crierie)

Chapter 1 Secret US Proposals of the Cold War

This angle shows to better advantage the pair of three-bladed contra-rotating propellers. Also note the added "bump" on the ventral fin to protect the props on rotation or landing. (John Adaz collection)

Behind the twin canopies are black slots representing the engine exhaust ports. Long drive shafts were used to turn the contra-rotating propellers which had an explosive device to blow them away in case of a bailout. (John Adaz collection)

Models and Artwork　　Chapter 1

In-house model of the XB-43 shows the jet converted airframe of the XB-42 now housing the two General Electric TG-180s. These two axial flow turbojets replaced the two Allison propeller engines. Without contra-rotating propellers there was no need for the ventral fin with the protective bump. As compensation a much taller vertical stabilizer was added. (John Adaz collection)

This angle shows one of the flush air intakes on the XB-43's left fuselage for the turbojets. The XB-42's aerodynamically clean airframe made it well suited for turbojet conversion, so the Air Force cancelled the Mixmaster program and ordered two prototypes of the XB-43 Jetmaster. (John Adaz collection)

Above: *Long twin exhaust pipes were need on the XB-43 for the General Electric TG-180 axial flow turbojets. Locating the jet exhaust side by side allowed for better control in case one turbojet went out.* (John Adaz collection)

American four-engine B-45 Tornado, an aircraft designed from the beginning as a jet powered bomber.

The one surviving XB-43 and the XB-42 were given to the National Air and Space Museum and serve as examples of a prop design that transitioned into a jet. It is rare to find the complete set of existing prototypes, their models, and artwork still surviving to this day.

Conceptual and Graphic Art

The use of art to manipulate emotions is a well-used tool in the advertising industry. In advertising it is the job of the ad agency to provide visual "comps" to their client, images of what their product would look like in magazine ads, billboards, web pages, and even TV commercials.

These illustrations are for the client to see and approve before producing the finished advertisement.

The relationship between the ad agency and the client is very similar to the relationship between aerospace and the military, except for one thing: the military is the client and the ultimate consumer. Aerospace industry relies on visual presentations to help sell their product to the military.

Traditionally, in a proposal, artwork is included along with the data specifications. The visuals range from simple three-views, graphs and charts, to photorealistic conceptual drawings and paintings, and now with computers, realistic 3D graphics and animation of the proposed aircraft can be presented. The Douglas XB-42 proposal brochure was made before the time of computers but it still relied heavily on the persuasive effects of conceptual and graphic artwork to tell a good story in order to sell the military on this new bomber.

The XB-42 graphics are used to visually compare its versatility as a medium bomber with a heavy bomber. It shows the amazing advantages it has over a "modern day" heavy bomber.

A two-page spread shows a sweeping arrow over the Pacific hemisphere from China to the island of Atu superimposed over Japan. This quickly conveys the idea that the XB-42 can cover the same range as the Boeing B-29.

The other two-page graphic literally compares the XB-42 to some sort of heavy bomber that appears to be a cross between the XB-19 and the B-32. At a glance one can see the XB-42's tremendous savings in terms of mission effectiveness, maintenance, and cost.

The conceptual art of the XB-42 is even more dramatic and reflects the sentiment of pay-back for the Japanese sneak attack on Pearl Harbor. The strafing illustration is of particular interest. The artist composed the destroyed Japanese Kawasaki Ki-61 *Hien* fighters in the foreground to look very similar to the iconic pictures of the destroyed Curtiss P-40s in the Pearl Harbor attack. Its intent was simple; to touch the heartstrings of the military men that considered buying the new Douglas design.

Illustration from the Douglas proposal brochure showing a formation of B-42s on a bombing run over some coastal industrial target while anti-aircraft shells explode around them. (National Archives via Ryan Crierie)

Illustration from the proposal brochure demonstrating the versatility of the B-42, now shown on a torpedo run against a heavily defended battleship. One torpedo has already found its mark and the other two are yet to come. (National Archives via Ryan Crierie)

The incredible range of the proposed B-42 bomber is touted in this illustration from the Douglas brochure. (National Archives via Ryan Crierie)

Illustration from the Douglas brochure showing the strafing capabilities of the XB-42 after it has completed the bombing run over a Japanese airfield. It is ironic that this design was being sold to the Air Force yet the illustrator was a navy man, later known as Admiral Arthur C. Beaumont. (National Archives via Ryan Crierie)

Models and Artwork | Chapter 1

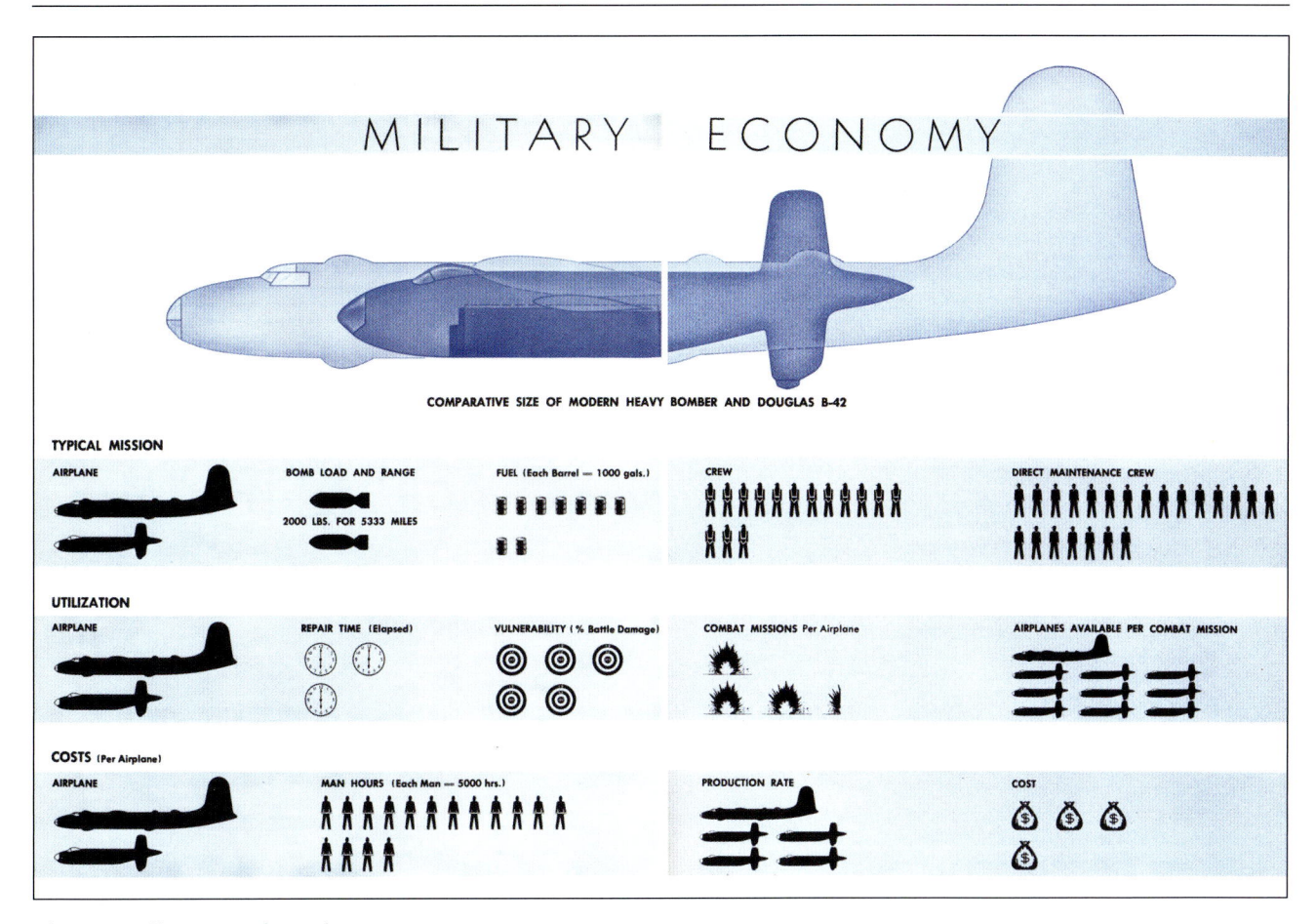

Above: *An illustration from the Douglas brochure comparing the B-42's economical advantage with a 'current' bomber that looks like a cross between the XB-19 and a B-32.*
(National Archives via Ryan Crierie)

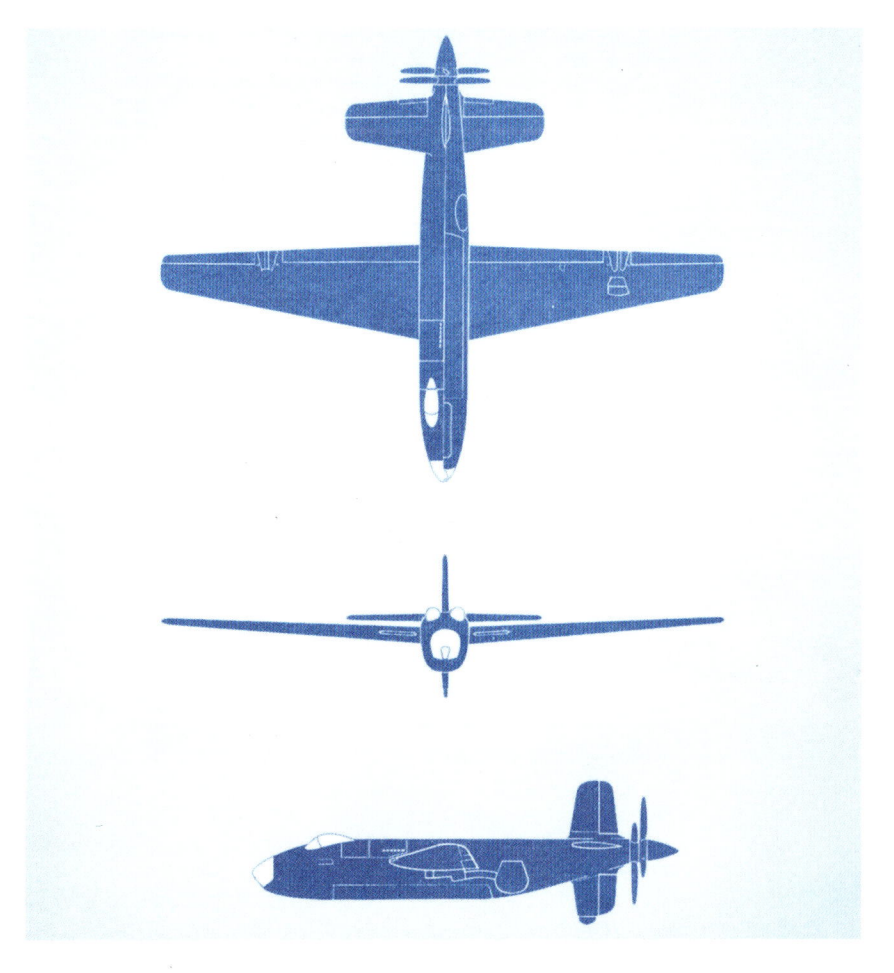

A three view of the XB-42 from the Douglas brochure.
(National Archives via Ryan Crierie)

25

Models During World War II

Long before World War II started, the model airplane industry had taken root in the United States. Men, boys, and even some girls became involved in the pastime of building and flying model airplanes. Such historical events as World War I, speed-record-breaking air races and Lindbergh's trans-Atlantic flight inspired many to start building and flying model airplanes.

During the 1920s and 1930s many model airplane companies sprang from small cottage industries; Ideal, Cleveland Model and Supply, Megow, Comet Models, Strombeck-Becker, and Peerless were companies manufacturing model kits, just to name a few. Their products ranged from scale to non-scale airplanes, stick-and-tissue flying models to solid wood models for display.

In early 1942, as America geared up for World War II, the Navy's Bureau of Aeronautics (BuAer) ordered the mass production of 1/72 scale plastic recognition models through Cruver of Chicago. These solid plastic model airplanes were to train soldiers and civilians alike in identifying aircraft as "Friend or Foe." To get immediate results while awaiting Cruver to start production, the Navy initiated the nationwide High School Model Building Program.

American model companies like Megow, Comet, and Strombeck-Becker supplied the schools with 1/72-scale solid wood model kits, plans, and templates. Having enthusiastic high-school kids carving blocks and slabs of wood into recognizable aircraft was a good way for the Navy to get the models quickly. Once Cruver started

Left: Most solid model kits were nothing more than a box of wood blocks needing to be carved into shape. Strombeck-Becker got a jump on the competition by offering models with pre-shaped pinewood parts that could be quickly put together and roughly resemble the plane. Here is what their Y1B-17 looks like right out of the box. Due to wartime material restrictions the usual metal propellers were substituted with cardboard. *(Author)*

Below: The same Strombeck-Becker Y1B-17 as above but with a little more detail and effort in sanding and painting. Strombeck-Becker, later to be known simply as Strombecker, would go on to help the war effort by offering a series of pre-shaped solid wood kits called the Spotter Models. The Spotter series made in 1/72 scale included a B-26C, P-40, SBC-4, A-20A, B-24D and a more modern B-17E. *(Author)*

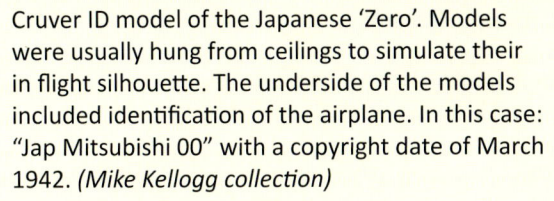

Cruver ID model of the Japanese 'Zero'. Models were usually hung from ceilings to simulate their in flight silhouette. The underside of the models included identification of the airplane. In this case: "Jap Mitsubishi 00" with a copyright date of March 1942. *(Mike Kellogg collection)*

Cruver P-51 Mustang. All the Cruver WW II ID models were uniformly made in 1/72 scale and were manufactured with injection-molded cellulose acetate, an unstable material which over time caused many of the airplanes to deform and crumble. *(Mike Kellogg collection)*

delivering its plastic models, the Navy decided to stop the High School Model Building Program, but reversed the decision, figuring that it would be better for young boys to already have some knowledge of aircraft identification, for soon these eager builders would be drafted.

On the Axis side of the conflict, the Germans made ID models out of both plastic and wood. The plastic models came in 1/200 scale and the wood ones were in 1/50 scale. Little is known about the wood ID models. They are black in color with the name of the aircraft imprinted under the wing. It is believed they were made in Czechoslovakia.

The Wiking-Modelle Story

Wiking-Modelle, headed up by Friedrich Pelzer, started producing 1/200-scale model aircraft in 1934. Made of metal, they were accurate representations of some ten different German aircraft. It was hardly surprising, therefore, that as hostilities approached, the German armed forces turned to the company to produce recognition models. For this task, Wiking-Modelle decided to use a hard thermo-setting plastic, which proved to be highly durable. (The plastic used gave the models a characteristic ring when dropped onto a hard surface, and doesn't soften when dipped in hot water—two of the tests modern-day collectors use to identify genuine items).

The series eventually encompassed 146 different models, production only ending when Germany was overrun. They represented the main combat aircraft of Germany, the United States, Italy, France, Holland, Great Britain, and Poland. They were distributed in four different basic sets, each in a briefcase-style, lockable wooden box with an inner glass lid. Each model had a small hole on the underside to enable display on a wire stand.

The early production runs were made of a very dark green material, and were subsequently referred to as the "schwarze" series. This soon changed to a slightly lighter (though still quite dark) green, which was used as the standard until the closing days of the war. Cockpits and glazings were turquoise, and each plane carried decals with the appropriate national insignia. Additionally, a small number of models were factory-airbrush-finished in camouflage, for presentation to high-ranking officers. In the final stages of the war, the dark green plastic was supplemented by a variety of colors, presumably because of the shortage of dyes.

After the war, Wiking-Modelle quickly returned to producing 1/200-scale model aircraft, starting with the British Meteor and Vampire jets, and later producing very neat models of aircraft like the B-47 and Super Constellation. However, the company did not resume supplying the newly reconstituted German forces. This contract went to Hansa. The problem was apparently a conflict over unsettled bills by the Third Reich, which the new government felt disinclined to honor! Hansa continued to produce 1/200 scale models for the German forces up to the 1970s (though generally not to the accuracy or quality of its Wiking predecessors). Put together, the two series cover every significant warplane (World War II Pacific theatre excluded) from the 1930s biplanes to the F-16.

Chapter 1 Secret US Proposals of the Cold War

The German Wiking ID models were much smaller than the American ID models. A comparison photo of a Wiking 1/200 scale Blohm-Voss Ha-139, on the left, next to a Cruver 1/72-scale version of the same airplane, on the right, shows the difference. Also the plastic used by Wiking to produce their models was much more durable than the celluloid acetate used by Cruver. *(Author)*

Wiking ID models were distributed in sets in lockable wooden boxes, each with an inner glass lid. This made for ease of transport and prevented losses. An outline of each aircraft was shown beneath its respective position. Few of these boxes survived the war and are very rare to find. *(George Cox)*

28

Models and Artwork   Chapter 1

Most Wikings were never painted except for the canopies, which were done in an aqua color. Their overall color was the color of the plastic they were molded in. Rare among the Wikings are these very special camouflaged RAF editions that were presented only to high ranking German officers. *(George Cox)*

Today, Wiking models remain highly collectible with many extensive collections existing in private hands and museums. Their historic appeal is heightened by their appearance, the extensive range of aircraft represented, and the fact that they don't deteriorate. Many of the models are readily available and it is not difficult, or expensive, to start a collection. However, some of the later aircraft like the Me-262 are exceptionally hard to find, as are the three Polish aircraft in the range, which were only produced for a very short time. The camouflaged models are also rare, as are examples of the original schwarze series. And, inevitably, very few of the original cases have survived.

The Germans also produced a line of 1/50-scale ID models. They were made from wood, painted black with gray for the glass. This Do-217 appears to have been enhanced with newer markings. Like the Cruver models, the German ID models had the aircraft identification stamped on the underside of the wing. *(Author)*

Birth of Advanced Design

Advanced design is a relative term, relative to its time period. Since the beginning of aviation, entrepreneurs who built and flew the first airplanes were constantly experimenting, adjusting, and re-designing their inventions in an effort to improve their flying capabilities – advanced design.

This was an expensive enterprise. It took money to build these inventions, including their up-keep and the upgrades. In 1909 the Wright brothers were the first to find a steady customer, the United States Army Signal Corp. Selling their Wright Military Flyer established the first relationship between an aircraft manufacturer and the military, a relationship that grew slowly, but wars would intensify it.

World War I forced British and European aircraft makers to accelerate their engineering and airplane designs in competitive attempts to out-do the enemy. The Americans came late into that war and not having been directly affected, their aviation industry stayed relatively behind the rest of the world.

Between the great wars, the U.S. aviation industry gained some momentum, developing aircraft that more or less kept pace with the other countries.

But in the mid 1930s, Germany made their move, cranked up their military production, and by the end of World War II were at the forefront of advanced aviation designs and technology. These accomplishments would have a profound influence on the aerospace industry after the war.

When the United States entered WWII, it mass-produced as much equipment as necessary to win, flooding the battlefields with tanks, trucks, guns, ships, and airplanes, It became a matter of time before enemy forces would run out of their equipment that was being shot out from under them. The war was won through attrition although the quality of American aircraft made for the war was sufficiently matched to most of the enemy's propeller driven airplanes.

The overwhelming U.S. production of bomber aircraft was used to neutralize the enemy war industry when the American and British air forces started to fly day and night missions over Germany. Up to one thousand bombers would fly over enemy territory in one single operation.

The Germans countered with their advanced aircraft designs to defend the homeland, experimenting and building aircraft for the purpose of climbing quickly, intercepting, and shooting down enemy bombers. Interceptors were not a new concept, but Germany brought new meaning to the term *'interceptor mission.'* They developed superior propeller-driven aircraft, but more to the point, they also came up with jet powered and rocket-powered interceptors.

The Messerschmitt Me-262 Schwalbe (Swallow). This aircraft was the first turbojet to be used in combat. It had a maximum speed of 540 mph (869km/h) with a ceiling of 37,565 ft (11,449.8m). Tamiya 1/48-scale model. (Jim Copeland Model)

Models and Artwork   Chapter 1

These new designs were pushed into combat prematurely, but times were desperate. And when the first combat jets and rocket interceptors showed up in the skies, they gave the Allies a big scare and left a deep impression.

Imagine, for a moment, the plight of a North American P-51 Mustang pilot escorting B-17 and B-24 bombers over Germany, watching for enemy aircraft, confident that his steed can take on all comers, when suddenly, a Do-335 easily slides right by as if he were standing still. Or out of nowhere, an aircraft with no propellers! – the Me-262 – just whizzes over at great speed and starts firing into the bombers. Or maybe our fighter pilot sees in the distance a streak, the Me-163, punching through the bomber formation and coming back down with cannons blazing. The immediate response is to slam the throttle into combat power and pursue the enemy, but even at top speed our fighter pilot cannot even begin to catch up with any of them. Imagine the feeling of helplessness our Mustang pilot must have felt.

The Allies were worried because the Germans had raised the technological bar of advanced designs and were not sure how it might affect the war. Britain and the U.S. had been developing their own jet fighters, but Germany had beat them to the punch. Also, the Germans were not only flying rocket interceptors but also sending two kinds of unmanned jet/rocket powered bombs towards England as retaliation for the daily bombing. The Allies had no such weapons in their arsenal.

Besides the Messerschmitt Me-163 Komet, the Germans were working on another manned rocket-powered interceptor, the Bachem Ba-349 Natter (Viper). Non-manned jet or rocket-powered weapons included the V-1 Flying Bomb, and the V-2 missile, the first long-range ballistic missile to be used in combat. The V-2 could carry a 1-ton (907kg) warhead 50 miles (80km) up into the atmosphere and hundreds of miles downrange to its target. It epitomized where German advanced design was completely in a league of its own compared to the Allies. The scientists involved with the V-2 program would became the most highly sought-after prizes at war's end.

The Messerschmitt Me-163 Komet was the only rocket-powered fighter to have become operational in WWII. With only five minutes of fuel, its tactic was to rocket above the stream of enemy bombers, level off and dive down through the formation, firing 30mm cannon rounds, then glide back to the airfield. DML 1/48-scale model (Rick Kosta model)

Chapter 1 Secret US Proposals of the Cold War

For the Germans, advanced designs were too little, too late. As Allied troops began overrunning the German countryside, American, British, French, and Russian forces were given their first up-close look at these unusual jets and rockets; some abandoned in the fields, others left on the production lines, and many still on the drawing boards.

These discoveries created a mad scramble among the Allies at war's end. The Americans, British, and Soviets scoured the German cities and countryside in search of as many German rocket and jet scientists and engineers as they could find. Many of the unusual aircraft and documents were gathered, packaged and shipped back to the Allies' home countries to be analyzed and used by their aircraft manufacturers.

One unusual design discovered in the field were the Horten Ho 229 flying wings, developed by the Horten brothers. The Ho 229V-1 was a test glider and the V-2 and 3 were turbojet powered aircraft. The Ho 229 became the first turbojet powered flying wing ordered into wartime production although Northop was already constructing their piston-powered XB-35s. After the war the V-3 along with other Horten gliders were sent back to the U.S. and Northrop did receive the gliders for their own use.

The Junkers Ju-287, a resin model in 1/48-scale by Al Parker was designed as a high-speed heavy bomber. This forward-swept wing prototype had been completed and flown in August 1944. The Soviets captured it and took it back to Ramenskoje for further testing. (John Aldaz collection)

The Horten Ho 229 was the first German flying wing to reach production status before the end of hostilities stopped any further progress. Powered by two Junkers Jumo 109-004B turbojets, the Ho 229V-3 was captured and shipped off to the US for further research. The other five Ho 229 airframes were destroyed to avoid having them fall into Soviet hands. A Dragon 1/48-scale plastic model (Allen Hess model)

Models and Artwork　　　　　　　　　　　　　　　　　　　　　　　　　　　　　　　　　　　Chapter 1

The Messerschmitt P.1101 V1, a 1/72-scale plastic model by Huma. Just days before war's end, American troops discovered an 80%-complete Me P.1101 V1 whose adjustable wings could be swept from 35 to 45 degrees, although only on the ground between flights. Bell Aircraft Co. engineer Robert J. Wood was sent to assess the jet, which was shipped to Bell in Buffalo, New York. From this, the X-5 was developed – the first jet to fly with variable-geometry wings. (Jim Copeland model)

What it may have looked like if the Focke-Wulf Ta-183 Huckebein, were dressed in combat colors. The designs for the Ta-183 were captured by the Soviets at the end of the war but it was never built. Several years later when the MiG-15 'Fagot' came out, it was suspected to have been inspired by Ta-183 documentation. They may seem similar, but in fact are quite different. AMTech Models 1/48-scale. (Jim Copeland model)

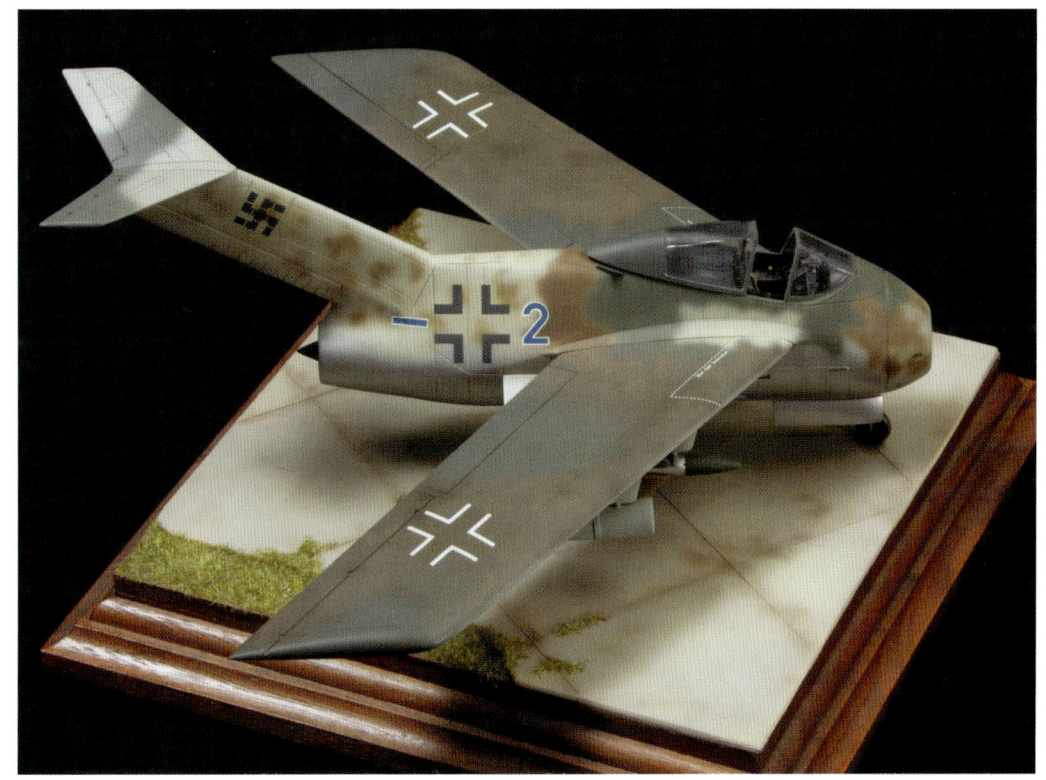

33

Three of the Horten gliders, two Ho IIIs and one Ho VI did arrive at the Northrop Hawthorne plant. The only surviving Ho 229 got as far as Freeman Field in Indiana, then languished in Park Ridge, Illinois before it was sent to the Smithsonian's Silver Hill facility in 1952.

Two other unusual aircraft discovered in the field were the Lippisch prototype DM-1 glider and the Messerschmitt P.1101 jet, a variable-geometry swept-wing aircraft. Both were found in unfinished conditions.

The Lippisch prototype DM-1 glider delta wing was to be powered by a ramjet. Eventually its inventor, Alexander Lippisch, was brought to the U.S. after the war and consulted with Convair on that company's advanced delta-wing designs. Convair's first delta prototype, the XF-92A would subsequently evolve into the advanced interceptors, the F-102 Delta Dagger and the F-106 Delta Dart.

The Messerschmitt P.1101 was an experimental aircraft used to test adjustable swept wings. However the different angled wings had to be manually changed before each flight. After the war, this aircraft was given to the Bell Aircraft Company which went on to produced a similar looking aircraft, the Bell X-5. But the X-5 featured mechanically swept wings that could change their sweep angle during flight.

The Germans had pushed the envelope in advance designs and now the victors were reaping the spoils. But with the end of WW II came the beginning of the Cold War. That new war gave a major boost to developing advanced aircraft as quickly as possible; for there was now a need for bombers with the capacity of delivering the most lethal weapon to a faraway enemy and for fighters with the speed and ability to intercept an enemy attempting to deliver such a weapon.

The Cold War energized the relationship between the aircraft manufacturers and the military like never before. The military started requesting such outrageously advanced designs that the manufacturers, at first, could not meet the specifications. They had to invent new technologies in order to satisfy the demands. The American aerospace industry now entered a prolific era of dreaming, designing, and proposing aircraft for the defense of the United States. And much like the Germans, American aerospace created spectacular advanced designs. Some actually became production aircraft, a few flew as prototypes, while others never made it past the mock-up stage, and many only existed as drawings and, of course, as models.

This Fiesler Fi-103, also known as the V-1 'Buzz Bomb', is an Al Parker Model. The first, somewhat successful cruise missile, the V-1 was launched from France and with a crude guidance system, could reach London. The Nazi Propaganda Ministry called it the Vergeltungswaffe *"retaliation weapon". (John Adaz collection)*

Models and Artwork | Chapter 1

Lippisch P.13a, plastic 1/48-scale by Dragon Model of the Supersonic Delta Wing Fighter. This proposed ram jet version, never built, was based on the prototype DM-1 glider. The unfinished DM-1 glider was discovered by American troops. Lippisch was ordered to complete it, then it was packed and shipped to the U. S. where it was tested in a wind tunnel. Eventually these studies helped Convair develop their delta program for the XF-92 and F-102. (Allen Hess model)

The Focke-Wulf Triebflügel was patented in 1938 but design work didn't start until 1944. The three wings rotated around the fuselage acting like a giant propeller. At the end of each wing was a Pabst ramjet that could be started with Walter rocket engines fitted inside each ramjet pod. Although the Triebflügel was never built, a wind tunnel model was tested up to speeds of Mach 0.9. This is a plastic 1/72-scale model by Huma. (Jim Copeland model)

Chapter 2
The Bombers

AS THE POPULAR AVIATION SAYING GOES: "Fighter pilots make movies and bomber pilots make history." While that point may continue to be argued for years to come, it is an undeniable fact that the deadly payloads carried to their targets by large manned bombers have indeed changed the world in which we live.

Wartime Development

Fighters have certainly had their strategic impact. Never more so than in the Battle of Britain where the heroic performance of the RAF's Spitfires and Hurricanes thwarted the threatened invasion of the British Isles by

This impressive large wooden model of the XB-36 measures about 36 inches (91cm) in wingspan. Prepared by Consolidated Vultee Aircraft Corporation (Convair), it sits atop a heavy globe base that signifies this aircraft is an intercontinental bomber capable of reaching anywhere in the world. Note the early red and white stripes on the rudders. (San Diego Air and Space Museum)

the German forces. Nonetheless, it was Japanese bombers that brought the USA into the war with the attack on Pearl Harbor, and two single American bombers that brought the war to a close four years later.

The dropping of the atomic bombs on Hiroshima and Nagasaki ushered in a new era. With the parting of the ways between former allies, the next four decades were to see an uneasy and often tense, stand-off between the West and the Soviet bloc. Initially the United States was the sole possessor of nuclear weapons but this advantage could not last and Stalin announced to the world in 1949 that the Soviets now had their own

The Bombers Chapter 2

atomic bomb. The ideological war was now backed with a terrifying capability.

Now that both sides had the bomb, urgent attention turned to a shared problem: neither possessed a bomber capable of flying far enough with its heavy, deadly cargo, penetrating enemy defenses and returning home safely. With the technology that existed during the previous war, flying from England to Germany or Tinian to Japan took only a matter of hours. To reach halfway around the globe from the United States to the Soviet Union required the better part of a day, and in 1946, there were no bombers that could fly either way between the Soviet Union and America.

By the time the Consolidated XB-36 made its first flight in 1946, the newly created Strategic Air Command considered it only as a stopgap measure. Instead, focus shifted to manufacturers engaged in harnessing new and improved technology for production of a new jet bomber.

With its 10,000-mile (16,093km) range, the B-36 gave America its first true "intercontinental" bomber. It could fly to Russia at high enough altitudes to avoid existing enemy fighters and then return safely to its base. But the military knew it was only a matter of time before this slow-moving target with no escort protection would eventually fall prey to advanced jet-powered enemy interceptors.

Meanwhile the head of Strategic Air Command (SAC), General Curtis LeMay, would have to rely on whatever he had in the inventory to give the sense that

This rear view of the XB-36 model shows the placement of the two top turrets, the dorsal turret and the tail gun. This was the defensive thinking of the Second World War. (San Diego Air and Space Museum)

America was strong, willing, and capable of protecting its citizens, defensively and offensively. It was during this time that airplane manufacturers stepped up their "out-of-the-box" creative thinking to come up with ideas and aircraft that could push the limits of technology to where the military wanted it to go.

Some of this creative thinking resulted in altogether new concepts. But it also incorporated examination and refitting of existing aircraft to improve performance. Observing the lifespan of the B-36 program, one can see new technology was progressively incorporated into the B-36. While it could not easily be converted to all-jet power, there was no problem in hanging an extra four jet engines beneath its massive wings. It thus

This plastic or fiberglass Convair B-60 model has single jet engines on each pylon under the wing. The final version had eight engines. It is speculated that this is an early variant of the B-60. (Jonathan Rigutto collection)

37

Chapter 2 — Secret US Proposals of the Cold War

A Planet Models 1/72-scale resin model of the early Northrop N-1M 'Jeep' flying wing. This model was the first true flying wing produced by Jack Northrop. Notice the pronounced anhedral on the outer wing panels. Later variants were given straight wings. (Barry Webb model)

became a ten-engine aircraft – "Six turning, four burning," as its pilots would report. This latter version was re-designated as the B-36D.

Before the war started, Jack Northrop had been working on flying wings and felt that this type of aircraft could actually be made into a bomber. His experiments with the N-1M in 1940 proved to the Army Air Corps that the all-wing concept could work. The N-1M "Jeep" became the first American flying-wing aircraft.

When the military issued the Request For Proposals (RFP) in 1941 for the 10,000 mile (16,093km), 10,000-pound (4536kg) bomb load intercontinental bomber, Northrop put up his design of the XB-35 and Convair offered its Model 35 (later designated the XB-36). Both proposals were accepted. The N-9M was built at about 1/3 the size of the XB-35 as a scale model to test out the control surface configuration and explore other potential problems. The logistical data gathered from hours of

This view of a Northrop fiberglass model of the XB-35 shows the bubble canopy and leading edge co-pilot and observation windows. Also seen are the leading edge air intakes. In the rear are the four Pratt & Whitney R-4360 Wasp Major engines driving a pair of four blade contra-rotating propellers on each engine. (Greg Barbiera collection)

flying the N-9M were incorporated into the larger flying wing version, the XB-35.

In November 1941 the Army Air Corps ordered the development of the first XB-35. Eventually the contract called for two XB-35s and thirteen YB-35s, however, only six wings were completed and flown. The first version of the XB-35 was equipped with four Pratt & Whitney R-4360 Wasp Major engines, each driving a pair of four-blade contra-rotating propeller. Northrop issued the following statistics regarding the specifications and performance data of the flying wing:

Wingspan	172ft (53m)
Length	53ft 1in (16.2m)
Empty weight	89,300 lb (40,590 kg)
Maximum takeoff weight	209,000lbs (95,000kg)
Loaded weight	180,000 lb (82,000 kg)
Maximum speed	393kts (632km/h)
Service ceiling	39,700 ft (12,100 m)
Range	8,150 mi (13,100 km)

Compare that to the Consolidated XB-36 statistics which also used the same Pratt & Whitney R-4360 engines:

Wingspan	230 ft 0 in (70.12 m)
Length	162 ft 1 in (49.42 m)
Empty weight	166,165 lb (75,530 kg)
Maximum takeoff weight	410,000 lb (186,000 kg)
Loaded weight	262,500 lb (119,318 kg)
Maximum speed	418 mph (363 knots, 672 km/h)
Service ceiling	43,600 ft (13,300 m)
Range	10,000 mi (16,000 km)

The XB-36 appeared to be the superior aircraft, but given enough development time, the YB-35 would actually out-perform the B-36.

Unfortunately the XB-35 and the YB-35 were plagued with problems during their development. The contra-rotating propeller suffered from inadequate gearboxes and governors causing excessive shaft vibration. When the propulsion system was switched over to conventional propellers, speed and performance dropped.

Also, there were CG (center of gravity) issues resulting in stability problems. Additionally, because of the shallow depth of the wing it could not accommodate the single large bomb bay necessary to carry an atomic bomb.

The YB-49 used the airframe from the XB-35 with integrated turbojet engines in place of the Pratt & Whitney R-4360 Wasp Majors. Four vertical stabilizers and fence wings were added to the airframe to improve lateral control. (Allen Hess model)

Chapter 2   Secret US Proposals of the Cold War

Above: A scratch-built 1/48-scale model of the Northrop YB-49 is shown with a same-scale plastic model of the N-9M. The YB-49 is a multi-media model made primarily out of Styrofoam and fiberglass with small metal and plastic parts. The N-9M is a 1/48-scale Sword model. N-9Ms were one-third the size of the large flying wing and became trainers for future XB-35 and YB-49 pilots. (Allen Hess model)

Left: In-house Northrop fiberglass model of the YB-49. This top view shows that the YB-49 was essentially the same airframe as the XB-35, but a little cleaner aerodynamically. However, the vertical fins and wing fences were not enough to give the stability desired in a bombing platform without an automatic flight control system – of which only one aircraft had installed to good effect. (Author)

Teething problems caused the program to fall seriously behind schedule, but first flight of the XB-35 came in June 1946. But what may well have been the biggest hindrance to the X/YB-35 was its design, simply too unconventional. In the end the military seemed more comfortable choosing a more conventional design, the XB-36 over Northrop's wing.

However, the Army Air Corp gave Northrop a second chance by requesting an all-jet version of the wing: the YB-49. This new wing would now be reclassified as a medium bomber and a reconnaissance platform if it went into production.

By adding eight GE/Allison J35-A-5 turbojets to former YB-35 airframes, the YB-49 flew faster and higher but at a greatly reduced range and brought in new aerodynamic problems. The YB-49 experienced phugoid oscillations (the Dutch-roll variant) due to the absence of the lateral stability formerly provided by the four R-4360 external shaft housings of the YB-35s. Although the YB-49 had been given wing fences and four small vertical fins, they were less effective in providing that stability.

In an attempt to keep the flying wing program alive, Northrop offered alternative versions in bomber and cargo configurations which were illustrated in a proposal brochure. These new wings would revert back to contra-rotating propellers using either two Turbodyne V engines (a projected version of the Northrop-Hendy XT37 turbo-prop engine) or four Allison XT40 turboprops. They had a longer "fuselage" that deepened its center section and extended the cockpit well in front of the leading edge. This gave it the depth it needed to accommodate the large atomic bomb of the time.

The Bombers   Chapter 2

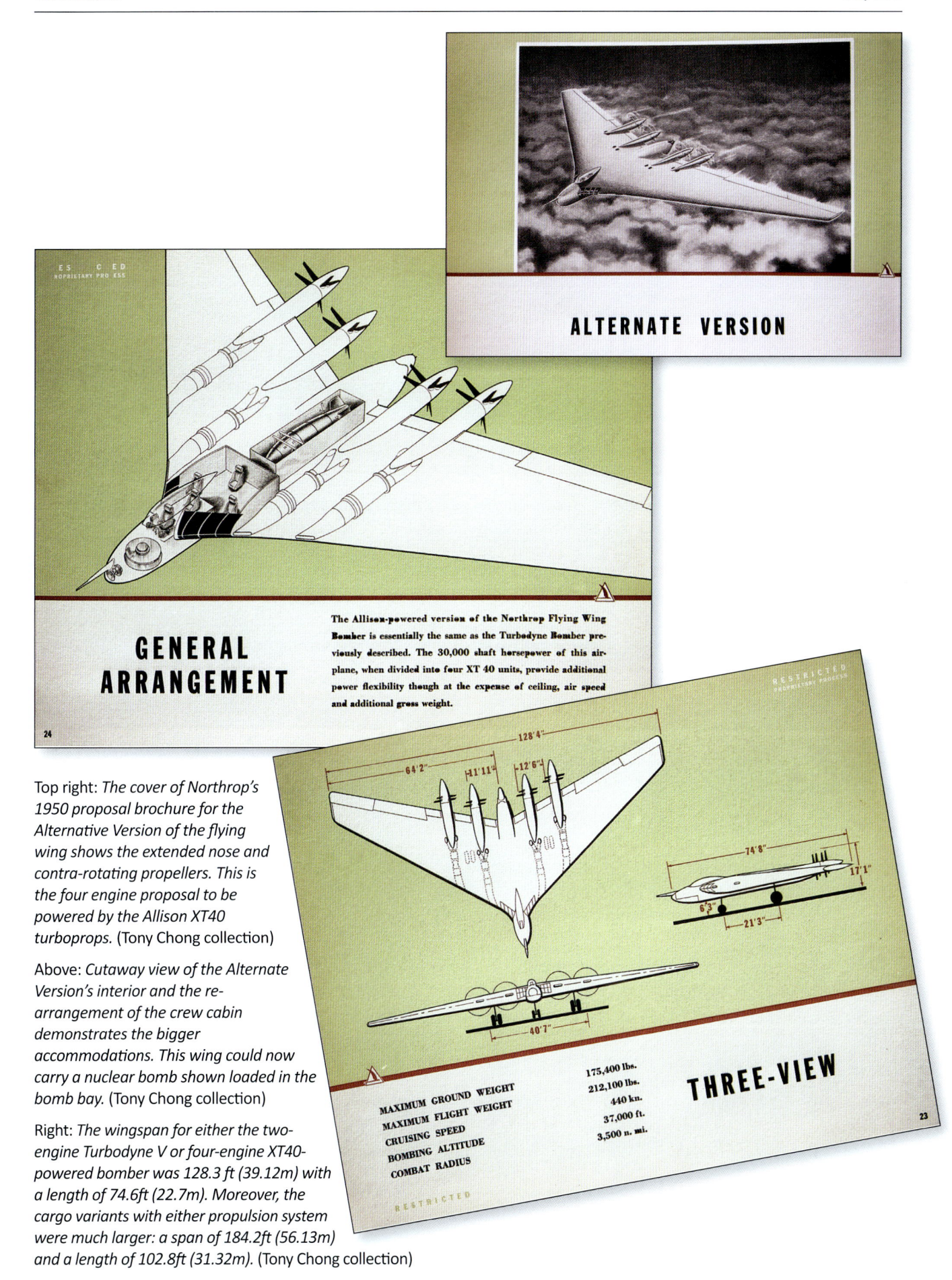

The Allison-powered version of the Northrop Flying Wing Bomber is essentially the same as the Turbodyne Bomber previously described. The 30,000 shaft horsepower of this airplane, when divided into four XT 40 units, provide additional power flexibility though at the expense of ceiling, air speed and additional gross weight.

ALTERNATE VERSION

GENERAL ARRANGEMENT

THREE-VIEW

MAXIMUM GROUND WEIGHT   175,400 lbs.
MAXIMUM FLIGHT WEIGHT   212,100 lbs.
CRUISING SPEED   440 kn.
BOMBING ALTITUDE   37,000 ft.
COMBAT RADIUS   3,500 n. mi.

Top right: *The cover of Northrop's 1950 proposal brochure for the Alternative Version of the flying wing shows the extended nose and contra-rotating propellers. This is the four engine proposal to be powered by the Allison XT40 turboprops.* (Tony Chong collection)

Above: *Cutaway view of the Alternate Version's interior and the re-arrangement of the crew cabin demonstrates the bigger accommodations. This wing could now carry a nuclear bomb shown loaded in the bomb bay.* (Tony Chong collection)

Right: *The wingspan for either the two-engine Turbodyne V or four-engine XT40-powered bomber was 128.3 ft (39.12m) with a length of 74.6ft (22.7m). Moreover, the cargo variants with either propulsion system were much larger: a span of 184.2ft (56.13m) and a length of 102.8ft (31.32m).* (Tony Chong collection)

Chapter 2 Secret US Proposals of the Cold War

Low-observable technologies along with its high aerodynamic efficiency and large payload capacity give the B-2 Stealth Bomber a significant advantage over any other aircraft. (Author)

But it was not enough, for the military cancelled the program in 1949 and by 1950 all airframes of the flying wing, except one YRB-49A, had either been destroyed in testing or scrapped. By December 1953 this last survivor was also put to the torch.

From its ashes the phoenix will rise and 35 years later the ultimate version of the flying wing, the B-2 Spirit, took to the skies. Its first flight was July 1989 and is considered the pinnacle of bomber and stealth technology.

If Jack Northop ever felt vindicated about his flying wings, it must have been at his old Hawthorne facility where he was briefed on the proposal phase of the ATB (Advanced Technology Bomber) around April 1980. He was 84. On 20 October 1981, eight months after his death, the Northrop Corporation won the contract to build the B-2.

This B-2 model was constructed from fiberglass by Northrop. The B-2's stealth properties come from a combination of reduced infrared, visual, acoustic and radar signatures, making it difficult to detect, track and engage the aircraft. (Author)

The Bombers

Composite Aircraft – Best of Both Worlds

The jet age had begun by the time World War II started. Both sides were developing their own designs, yet the Germans were first to fly jets into combat. By the end of the war the British were flying Meteors and the Americans had their P-80s. But jet propulsion was still a new technology and jet engines were lacking in efficiency and power.

The first American jet bombers were the Douglas XB-43 and the North American XB-45 Tornado. The XB-43 was the re-engineered airframe of the reciprocating pusher prop XB-42 and first flew on 17 May 1946. The Tornado was the first American bomber designed purely for jet power, and made its first flight on 17 March 1947. Its propulsion system was just adequate enough to get it up and flying, but for the most part, the military was disappointed with its performance. This was the era when engine manufacturers were all struggling to come up with a reliable and efficient turbojet with sufficient thrust to power a big airplane.

Some manufacturers decided to construct "interim" airframes, awaiting the arrival of the right turbojet. As a compromise, they constructed airframes that sought the best of both power-plant worlds: the fuel efficiency of the reciprocating engine or turboprop along with the speed and power of the turbojet engine. These were called composite aircraft. Thus a bomber could utilize jet engines to assist in takeoff and climb quickly to altitude; or during its mission use the ability to dash quickly in and out of the target area. And the reciprocating engine/turboprop would be useful for conserving fuel when flying to and from the target, loitering, and, finally, landing after its mission.

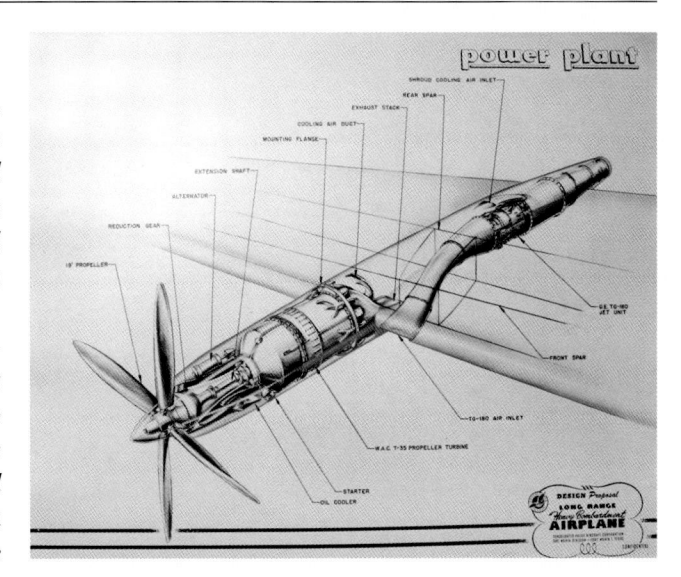

This cutaway illustration of Consolidated Vultee's Long Range Heavy Bomber concept shows a composite power-plant installation consisting of a Wright XT35 turboprop paired with a General Electric TG-180 turbojet. *(National Archives)*

Composite proposals were used in bomber and reconnaissance designs. Convair's eight-engine forward-swept wing Long Range Heavy Bomber came with four Wright XT35 turboprops plus four General Electric TG-180 jets housed in four nacelles. The turboprops were placed in the front of the nacelles while the jets were located in the rear.

The Douglas Model 1211, and Model 1240, both had the option of adding turbojets in the rear portion of the nacelles.

The following are a number of other composite proposals, all fairly conventional designs, offered as light bombers or reconnaissance aircraft.

Conceptual art of the North American RD-1401-1 gives the feeling of an aircraft that is all business; efficient and fast. The turboprops are spinning as the turbojets push the craft effortlessly through the air. *(National Archives via Ryan Crierie)*

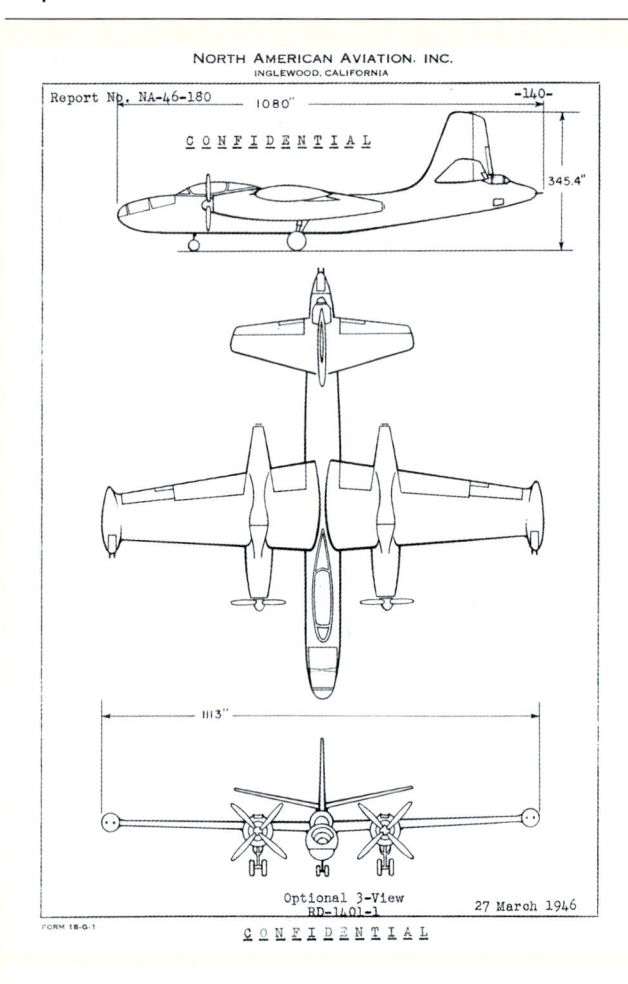

A three-view drawing of the North American RD-1401-1 dated March 1946. The length is given as 90ft (27.4m), the wingspan is almost 93ft (28.3m), and the height is almost 29ft (8.8m). The RD-1401 seems to be a beefed up version of the North American Savage that was ordered in June 1946. *(National Archives via Ryan Crierie)*

North American RD-1401-1 and NA-163 (XA2J-1)

The North American RD-1401-1 bomber was proposed on 27 March 1946. The RD-1401-1 looked like a prop version of the company's B-45 Tornado, but this was a larger aircraft. It carried a crew of five—a bombardier and navigator in the nose, two pilots in the upper fuselage, and the tail gunner seated at the rear. For protection, it had four two-gun turrets, two placed optionally on the sides of the forward fuselage or in wingtip nacelles, with the others underneath the rear fuselage and in the tail.

Below: Conceptual art of the North American XA2J-1 reveals a very sleek design with the elongated turboprop nacelles. The NACA-style inlet for the single turbojet placed in the rear can be seen on the top of the fuselage. *(National Archives via Ryan Crierie)*

The Bombers

The NA-163 was a second version of the North American Savage prototype. The XA2J-1 was Phase 2 of the strategic bomber program for the Navy. The BuAer asked North American for a turboprop-powered development of the Savage to be powered by two Allison T40-A-6s with four-blade contra-rotating propellers. The XA2J-1, which was known as the "Super Savage," was bigger and heavier than its predecessor. Its wingspan was 61 feet, 10 inches (18.8m) and its length was 54 feet, 11 inches (16.7m).

Douglas Model 1018

Somewhat similar in appearance to the North American 1401-1, the Douglas Model 1018 was powered by two T35 turboprops driving 19ft (5.8m)-diameter propellers and two J33 jets. This was proposed in April 1946 as either a bomber or an attack bomber. For the attack version four more fixed guns went in the nose, giving the craft 12 forward-firing nose guns. The Model 1018 had an estimated service ceiling of 40,000ft (12,192m) and a range of 820 miles (1320km).

There were other variants offered with some differences in engine configurations: The 1018-2 had dual-rotation gearing on its T35s with 16ft, 6in (5m) diameter contra-rotating Hamilton-Standard propellers. The 1018-5 was a jet-powered variant, with a larger wingspan of 106ft, 5in (32m) and a longer fuselage of 110ft, 1.6in (34m), making for an altogether sleeker design.

Lockheed L-238 Super Neptune

Lockheed began a study in 1953 to find the best configuration for the Navy's land-based ASW requirements. Over the course of three years, it came up with several iterations based on the Neptune P2V-7. The L-238-9 clearly exhibited much of the look of the company's Neptune anti-submarine aircraft but, in fact, was an entirely new aircraft for multi-mission capabilities and optimized with new power plants. It was nicknamed the "Super Neptune."

Two T54 turboprops were to drive contra-rotating propellers and there were two J71 jets also in the wingtip nacelles. The model has a detachable tail cone, containing twin remote-controlled cannons, which suggests that other options could have been available.

Yet the technology was still in its infancy. The reliability and efficiency of the turbojet had not reached a level of adequate performance. The turbojet had the potential to propel aircraft to higher speeds and altitudes than the reciprocating engine, but it was also much thirstier. One could fly very quickly for a short amount of time with a jet power plant, then run out of gas; not a desirable quality to have in any aircraft, combat or otherwise.

For the most part, these were short-lived experiments. Eventually, all of them were to become technological dead ends. The stop-gap solution would be overtaken by the advancement of turbojet technology.

Above: The Lockheed P2V-7 Neptune was a production aircraft that used composite propulsion systems. It is powered by two Wright R-3350-32W turbo compound engines and a pair of Westinghouse J34-WE-36 turbojet engines. Revell model. *(Craig Kodera)*

Left: A three-view of the Douglas DS-1018, which would have had a wingspan of 95ft (29m) and a length of 91ft 1in (27.8m). *(National Archives via Ryan Crierie)*

The Lockheed L-238 composite reconnaissance aircraft was a proposed larger version of the Neptune P2V-7. The model shows an elegant looking aircraft with its straight wings holding contra-rotating truboprops in the center and the jet pods at the wingtips. *(Author)*

Lockheed's in-house model of the L-238 is well crafted. Even the counter-rotating propellers work. When one propeller is spun the other turns in the opposite direction. Magnets built into the underside of the fuselage indicate optional pieces could be attached to show different configurations of radar, storage or fuel pods. *(Author)*

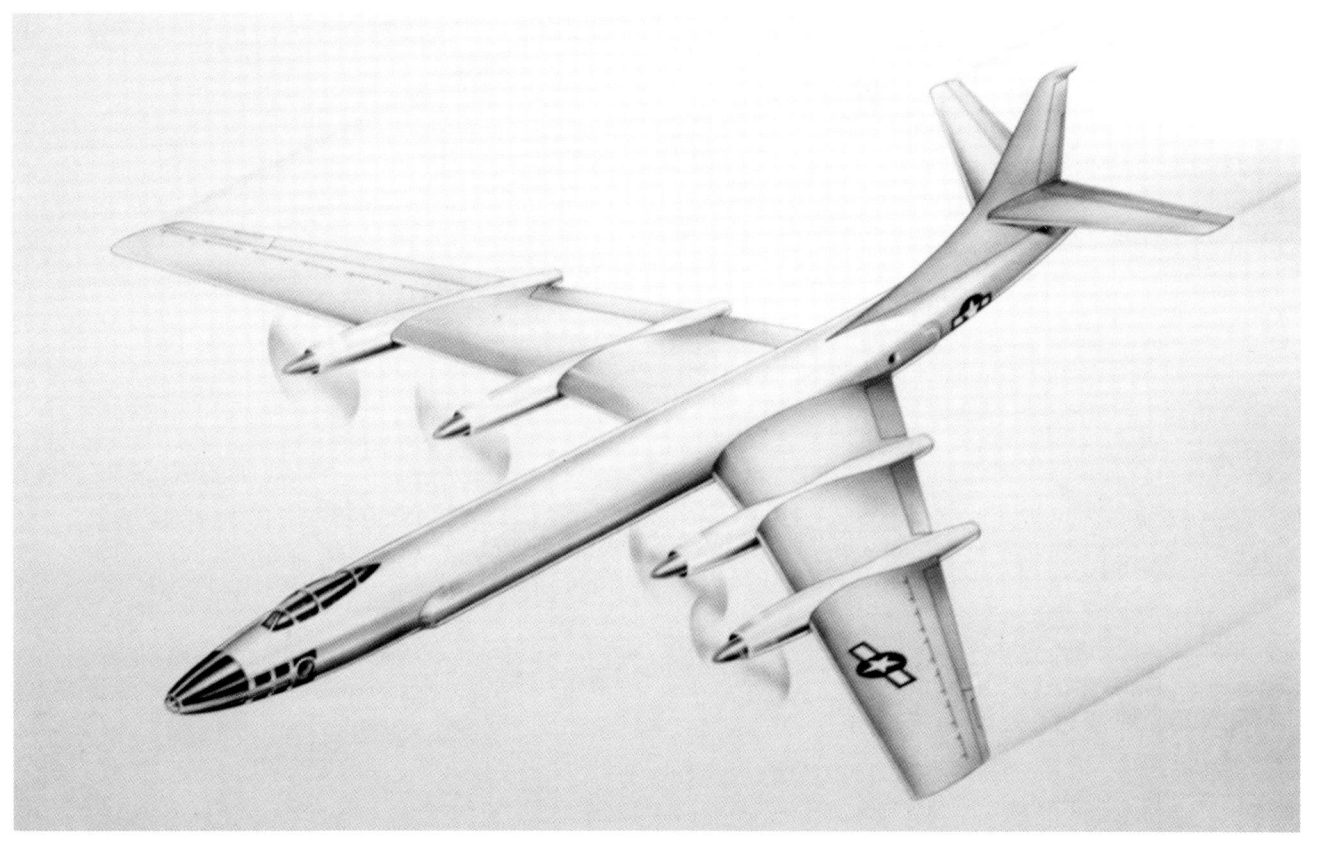

A proposal brochure illustration depicts the Convair Long-Range Heavy Bomber. This bomber had a 30-degree forward-swept wing and was powered by four Wright XT35 turboprops in the front and four General Electric TG-180s in the rear section of the nacelles. This was Convair's proposal to replace the B-36. (National Archives)

The Call For New Bombers

Even before World War II had ended, the Army Air Forces knew that jet bombers would be the future of military aviation. Although the XB-36 was under construction, the USAAF called for a number of new designs. During the war and just after, the Air Force issued a series of requirements and "characteristics" for different types: heavy bombers, medium bombers, and fighters that could also serve as attack bombers (attack aircraft were eventually known as light bombers). The aircraft manufacturers responded with their ideas and concepts either based upon existing projects or completely new designs.

The dilemma during this period was that the desired powerplants needed to permit these aircraft to meet the USAAF's requirements did not exist. There would thus be a lag period between making airframes capable of the required performance and the development of turbojets, or even turboprops, to power them.

By November 1944, the War Department had issued requirements to the aircraft manufacturers for a bomber that could reach 450 to 550mph (724 to 885km/h), fly at 40,000 to 45,000ft (12,192 to 13,716m), and have a tactical operating radius of 2,500 to 3,000 miles (4023 to 4828km). Although the specifications asked for in this document were simply beyond the current technology, the manufacturers were encouraged to try and come close to what had been requested.

When Boeing, Convair, Martin, and North American submitted their proposals, the military decided to move as quickly as possible, doing away with the normal design competition procedure. Instead it opted to go straight to building each manufacturer's prototype and picking the best-performing aircraft that could be put into production quickly. That honor went to the first American pure jet bomber, the North American B-45 Tornado. (See the sidebar "America's First Jet Bomber.")

It was also agreed that if any of the other designs proved to be superior, but needed more time to develop, then it too could be ordered into production and given a revised time frame. The Boeing B-47 Stratojet fitted the latter criteria.

Some very interesting bombers had in fact been proposed toward the end of World War II and during the post–World War II era though they did not fly until after 1945.

Jet power was the goal for most of these future planes, bombers, or otherwise. But, as already explained, in 1945 jet power was going through its teething period and the manufacturers had to wait until a reliable and efficient turbine power plant had been developed. Not only did the available engines fall short in terms of power and fuel consumption but time between major overhauls was measured in just a few hours of flying. To make a reliable intercontinental bomber, a turboprop engine was seen as possibly the best solution, at least for the immediate future Such designs did not have the high-speed capabilities of pure jets, but with their greater efficiency they offered longer endurance and better range. A number of airplanes were therefore developed with the turboprop engine as their power source. Designs powered by turboprops, turboprop/turbojet combinations, and, of course, pure jet power as engine technology advanced, are all represented in this chapter.

As stated earlier, this book is not intended to be a complete picture. The following selected examples are some of the more unusual designs to be included alongside more conventional aircraft for historical perspective. The pathway is mainly chronological, although occasionally it will jump time to follow designs that subsequently transformed into later models or to bring up references to future connections.

Douglas Aircraft Company

Model 1112

We have already described the propeller-driven Douglas XB-42 and jet-powered XB-43. Although Douglas did not participate in the original call for a long-range heavy bomber, it did design a couple of different models for possible consideration around 1946–1947. Models 1112A and 1112B were larger and sleeker variations of the XB-42, XB-42A, and the XB-43. Just like these predecessors, the Model 1112A had the same characteristics, consisting of contra-rotating propeller engines, or the prop, with added under-wing jets. The Model 1112B was the a pure-jet version. The distinctive Douglas "bug eye" canopies for the pilot and co-pilot were even retained. Wingspan for the Model 1112A was nearly 119ft (36.3m); for the 1112B, just over 138ft (42.1m).

Model 1155

The Model 1155 Douglas proposal dates from April 1948 when the design was described as an "Interim Strategic Bomber," yet a closer look at the model reveals similarities in the fuselage and wings of the DC-6 airliner. In fact, the Model 1155 was a beefed-up, jet-powered version of the DC-6. The wings were extended and the fuselage lengthened. A "bomber"-type nose, with extra glass for the navigator and bombardier, was added and a tail gun emplacement was located in the rear fuselage. The platform held the tail gunner had an almost a B-17-like aspect to it.

Model 1155 came in two versions: with four jet engines or with six jet engines. What is interesting to note is that the inboard engines on the six-engine model had side air inlets along the larger pod, while the outboard engines had apertures in the front of their nacelles. These inner pods were larger because they housed the main landing gear as well.

The glass nose displayed on this in-house wood model of the Douglas Model 1155 makes this airliner derivative look more like a bomber. (Author)

The Bombers   Chapter 2

This wooden model of the Douglas 1155 shows the six jet engines mounted on straight wings and the old style bomber glass nose. Note the conventional air intakes for the middle and outboard engines, while the two inboard engines have their air intakes on the side of the nacelles. (Author)

The Douglas Model 1155 was considered an Interim Strategic Bomber. Essentially, this aircraft was a beefed-up DC-6/7 airliner with elongated wings and fuselage. The tail group also reveals the linage of the DC-6/7. (Author)

49

Consolidated Vultee Aircraft Corporation (Convair)

As with many U.S. Aerospace companies in the years following World War II, earlier organizational structures were modified into more modern corporate entities. Such was the case with Consolidated Vultee which became Convair (Consolidated Vultee Aircraft). As one of America's premier aircraft manufacturers, Convair would go on to design and build advanced fighters, bombers, flying boats, airliners, experimental seaplanes and vertical-takeoff aircraft, and even Intercontinental Ballistic Missiles (ICBMs).

XA-44/XB-53

In 1945/46 Convair was working on two proposals, the three-engine XA-44 attack bomber and the four-engine XB-46 high-altitude medium bomber. The costs for both were prohibitive, so Convair worked out an agreement with the USAAF to cut two of the three XB-46 prototypes in order to continue with the XA-44 project. Of the two bombers, the XA-44 was much more unconventional with its forward-swept wing. Remember, this is the time when straight wings were the norm and to have swept wings, either backward or forward, was pretty innovative indeed.

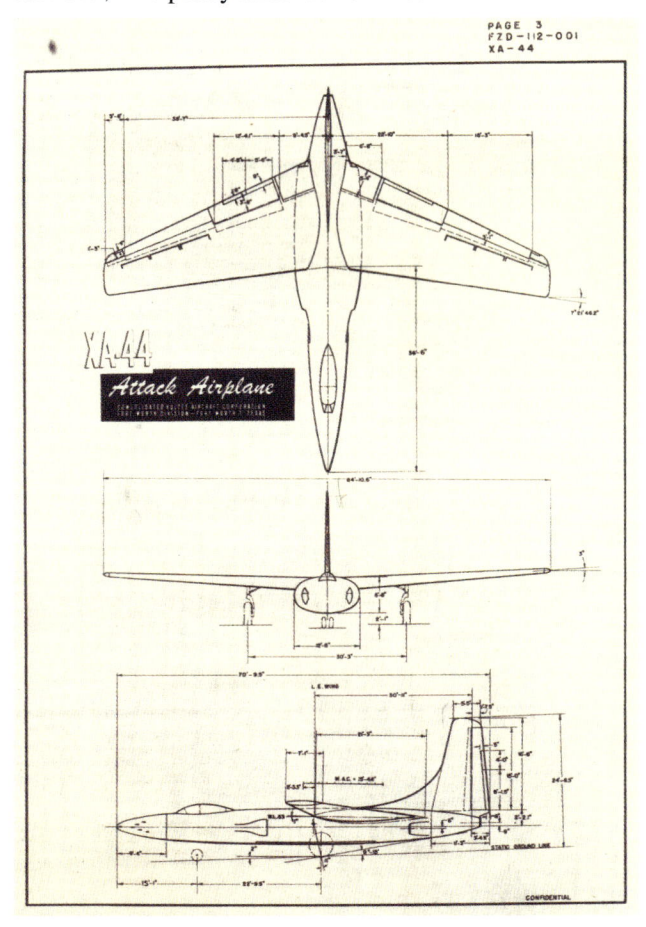

The Convair Model 112, was proposed in May 1945 and designated XA-44 in November. It was an aircraft directly influenced by captured World War II German documents, referencing the forward-swept-wing Junkers Ju-287 bomber as part of the windfall of German documents uncovered after the war was over. Jet flight was still pretty new in 1945 and Convair was trying to explore the advantages of forward-swept wings as well. This wing style eliminated tip stalls, had less drag, and had better handling of compressibility. It also enhanced low-speed flight for slower approach and landing speeds. The big problem that was never solved was the tendency for the wing to twist under aerodynamic pressures.

The attack bomber proposal had three General Electric TG-180 jets, all encased within the fuselage. Projected performance included the ability to: fly at 580mph (933km/h), reach 44,300ft (13,503m), and had a range of 1,500 miles (2414km). For ground support, it had 20 machine guns; 12 in the nose and the rest divided between two retractable turrets, one behind the canopy and the other underneath. It could also carry 12,000lb (5443kg) of bombs, two Mk13 torpedoes, or 40 HVARs (5-inch High Velocity Aerial Rockets). The XA-44 had a 30-degree forward-swept wing with 8-degree of dihedral. The large vertical tail had a rudder and the elevator function was on the inner trailing edges of the wings, while ailerons were on the outer portion.

By 1948 the newly established US Air Force had changed its aircraft designations and the "A" for attack category was eliminated. The XA-44 became the XB-53. The new bomber version was estimated to fly higher and have 500 miles (805km) of additional range. But the program was becoming too expensive, and was canceled before either of the two prototypes had been completed.

Long-Range Heavy Bomber

But the demise of the XA-44/XB-53 program brought out a bigger and better bird; the Convair Long-Range Heavy Bomber (LRHB). The Army Air Force was looking ahead to the next long-range heavy bomber to succeed the Convair XB-36, even before the Peacemaker was completed. So it issued in November 1945 its "Military Characteristics for Heavy Bombardment Aircraft" and in February 1946 the Air Technical Service Command (ATSC) released a Request for Proposals (RFP). Boeing, Martin, and Convair all submitted designs and Convair's

The Consolidated Vultee Model 112 was a high-speed attack airplane for the Army Air Forces, which was later designated as the XA-44. Consolidated Vultee was still considering the use of forward-swept wings with this proposal. (National Archives via Ryan Crierie)

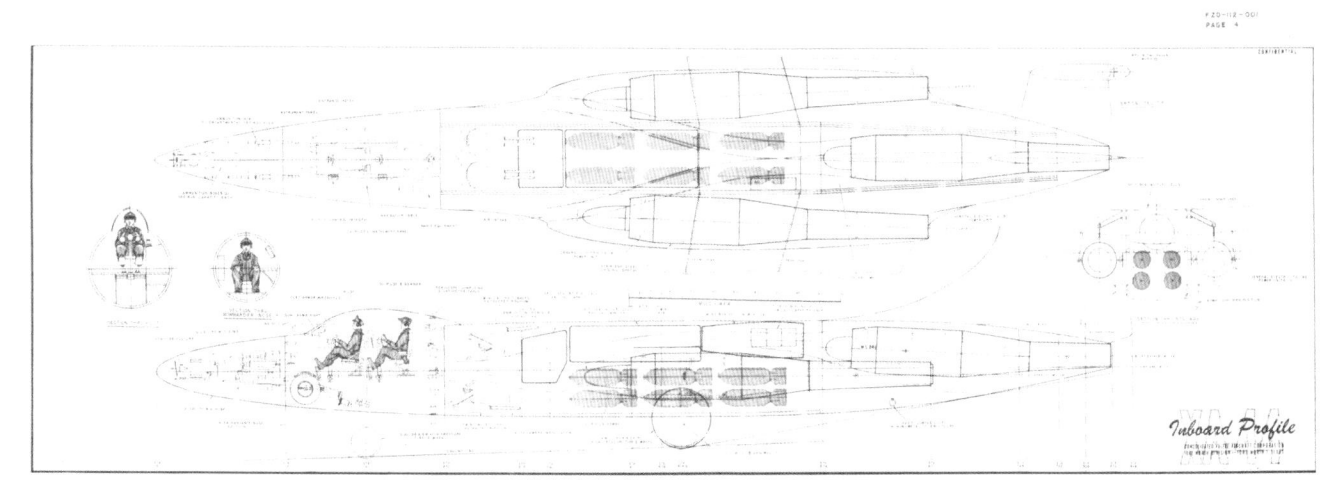

An illustration of the inboard profile for the Consolidated Vultee Model 112 shows the placement of crew, fuel tanks, bomb load, and the layout for the three jet engines. (National Archives via Ryan Crierie)

LRHB proposal represented its hope to continue the work it had started on forward-swept wing technology. (Also see the Martin 235 and the Boeing 462.)

The LRHB was not given any designation. It was a design proposal from 22 April 1946 for an interim bomber that Convair hoped to develop while awaiting bigger and better jet engines from future technology. Once the more powerful engines were produced, this bomber would become the long-range heavy bomber the USAAF was looking for. However, it did seem to be influenced by the work done on the XA-44/XB-53.

Like the XA-44/XB-53 this design had 30-degree forward-swept wings. Unlike the XA-44/XB-53 it had an elevator and its tail group was swept backward. The interim bomber was to be powered by four Wright XT35 turboprops in the front nacelles and four General Electric TG-180 turbojets in the aft end of the nacelles. The jets were used for takeoff and boost to speeds required during the mission.

With a wingspan of 167ft (51m) and length of just over 172ft (52.4m), this bomber could fly at altitudes of up to 44,300ft (13503m), with a cruise speed of 360mph (579km/h), a top speed of 520mph (837km/h), and a combat radius between 2,445 to 3,180 miles (3935 to 5118km). The overall gross weight was 235,000lb (106,594kg). For defense, the bomber had ten 20-mm guns. Four pairs were flush-mounted turrets, with two in the front-side fuselage and the other pair in the rear fuselage. There was one pair in the tail and all of the guns were remotely controlled

As noted, Convair was one of the aircraft companies that utilized the information captured from the Germans after World War II to develop its own innovative designs. The forward-swept wing technology was applied to the XA-44 and passed on to this larger, long-range heavy bomber. Convair also read up on the German World War II delta-wing experiments and actually consulted with Dr. Alexander Lippisch when it started proposing the delta-wing fighters, (discussed in Chapter Three).

In the end, the USAAF rejected the forward-swept wing design, just like the XA-44. It would take many years before a forward-swept wing aircraft would fly successfully, and that happened in 1964 when Hamburger Flugzeugbau introduced their forward-swept business jet, the HFB-320 Hansa Jet.

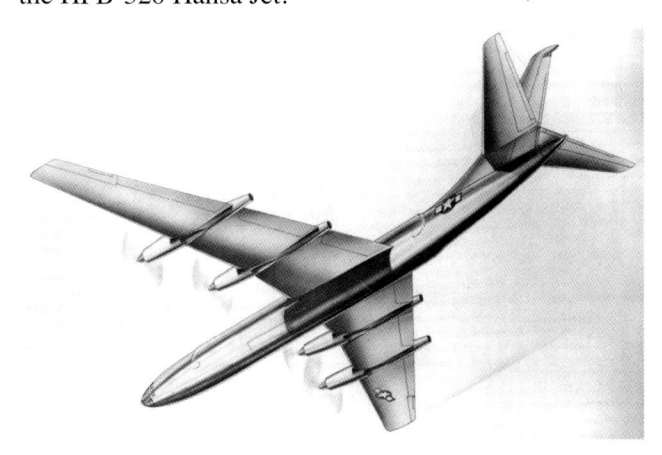

This view of the 1945 Consolidated Vultee Long-Range Heavy Bomber (LHRB) proposal shows its sleek lines, forward-swept wings and the flush rear turret just behind the wing. (National Archives via Ryan Crierie)

Republic Aviation Corporation

As a small company that came into prominence building more than 15,000 examples of the legendary P-47 Thunderbolt fighter bombers during World War II, Republic Aviation Corporation sought to expand its product line with any number of advanced designs in the Postwar era. These concepts and proposals included everything from a small general aviation seaplane to a massive four-engine, 450mph (724km/h) photo-reconnaissance aircraft. The following proposals came

Chapter 2 — Secret US Proposals of the Cold War

In-house Republic wood model of the AP-42 reveals some interesting details of the aircraft. The fuselage has a pointed glass nose and tail cone. The fighter-style canopy indicates the pilot and co-pilot/navigator would have been seated in tandem. The elongated nacelles under each wing house the engines and the main undercarriage. Note the small figure under the nose to indicate scale. (Cradle of Aviation Museum)

into being as the "little company that could" continued its relentless pursuit of growth and expansion.

AP-42

The Republic AP-42 currently has no supporting information. The unknown design shows Republic's characteristic tail group in V-tail, or butterfly, configuration. This tail design was used in its NP-50 Navy attack aircraft and also its AP-31 models that eventually became the XF-91 fighter. The tapered wings held long slender nacelles that housed not only the engine, but the main landing gear as well. It is speculated from the project number that this model dates from about 1947 and may have been part of the competition to replace the B-47.

Another view of the AP-42 shows Republic's characteristic butterfly or V-tail. The thin tapered, highly- swept back wings have leading-edge slats and trailing-edge flaps fitted on either side of the nacelles. It has a long slim fuselage with pointed glass nose and tail cone. (Cradle of Aviation Museum)

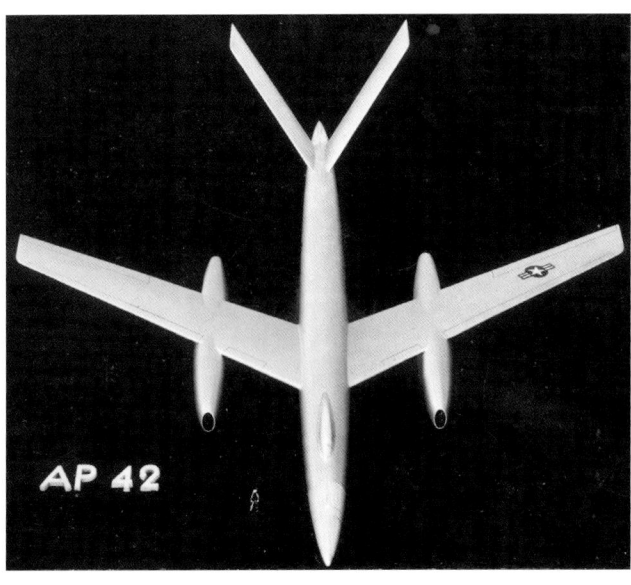

The Martin Company

If Republic was perceived as using advanced design proposals to expand and grow, it would be fair to say that the Glenn L. Martin Company was using its proposals to literally stay alive in the airplane-building business. Having produced a successful series of bombers, flying boats, and airliners, the challenges of the new jet age were proving insurmountable to Martin. Their last airplane would be an impressive four-engine jet-powered seaplane that showed huge promise but never entered production.

Model 236 Heavy Bombardment Airplane

On 23 April 1946, Martin proposed its "Heavy Bombardment Airplane" Model 236 in response to the ATSC February 1946 RFP based on the 1945 "Characteristics for Heavy Bombardment Aircraft." (See Convair's LRHB on page 50 and Boeing's Model 462 on page 54). Very little paperwork has been found on this proposal but it was discovered at the National Archives in an aerodynamic report, listed as Engineering Report No. 2349 which contained a three-view illustration. The document describes a conventional-looking bomber with an unusual defensive armament.

The aircraft was a six-engine, straight-wing heavy bomber. On its wingtips were large pods that housed eight 20-mm cannon and all of these guns could be remotely rotated in different directions. On each pod, the two cannon that were positioned to fire straight ahead had a lateral and vertical sweep of 90 degrees. The two that fired backward had only a lateral and vertical sweep of 80 degrees. The other four cannon that shot straight out to the side could swivel 130 degrees back and forth and rotate up and down 220 degrees.

The bomber was to be powered by six Wright Aeronautical Corporation XT35-1 gas turbines, driving six-blade dual-rotation Hamilton Standard Propellers. In the aerodynamic report the following data and performance statistics were listed:

Gross weight	275,000lbs (124,738kg)
For design, gross weight less half fuel load	221,145lbs (100,310kg)
Top speed	475mph (764km/h)
Tactical operating altitude	35,000ft (10,668m)
Service ceiling	41,700ft (12,710m)
Tactical operating radius	2,147 miles (3455km)
Design range	5,726 miles (9215km)
Average speed	407mph (655km/h)
Bomb load	10,000lb (4536kg)

Boeing Airplane Company

Not all of the strange designs and proposals by American aircraft manufacturers resulted in technological dead ends. Some of these unusual designs actually evolved, eventually becoming real aircraft. The Model 432 and Model 462 from Boeing are two examples of this evolutionary process.

As mentioned earlier, four aircraft manufacturers had submitted their proposals for the first post-war jet. The military had decided on the North American B-45 as the quick solution of producing the first jet bomber but after reviewing the other candidates, the USAAF elected to give Boeing extra time to develop its proposal.

Model 424

Boeing started with Model 424 in the beginning of 1944, a straight-wing medium bomber with large underwing engine pods. The engineers figured out that the large pods hampered the high-speed capability of the wing, so by December of 1944 they had followed up with Model 432.

Model 432 and Model 448

Model 432 was a straight-wing aircraft with four engines embedded in the upper-forward fuselage. By February 1945, the government had given Boeing a contract to develop a medium bomber, designated XB-47. Boeing also benefited from German World War II research–in fact captured documents about swept wings. It was concluded that a swept wing would increase the speed of the aircraft and Model 432 evolved into the Model 448, now with swept wings and six engines installed within the fuselage.

Model 450

The next iteration was the Model 450, where the engines were placed on pylons under the wing and the fuselage was trimmed down. The Model 450 now exhibited the appearance of the resulting B-47 Stratojet. And on 17 December 1947, the XB-47 flew for the first time.

This illustration of the six-engine Martin Model 236 Heavy Bombardment Airplane proposal is the only drawing of the aircraft found to date. It shows the wingtip pod that housed the eight 20mm cannon for defense, a 195ft (59.4m) wingspan and length of just alittle over 132ft (40.2m). Gross weight is listed as 275,000lbs (124,740kg). The engineering drawings show the various degrees of sweep each cannon has on the pod. (National Archives via Ryan Crierie)

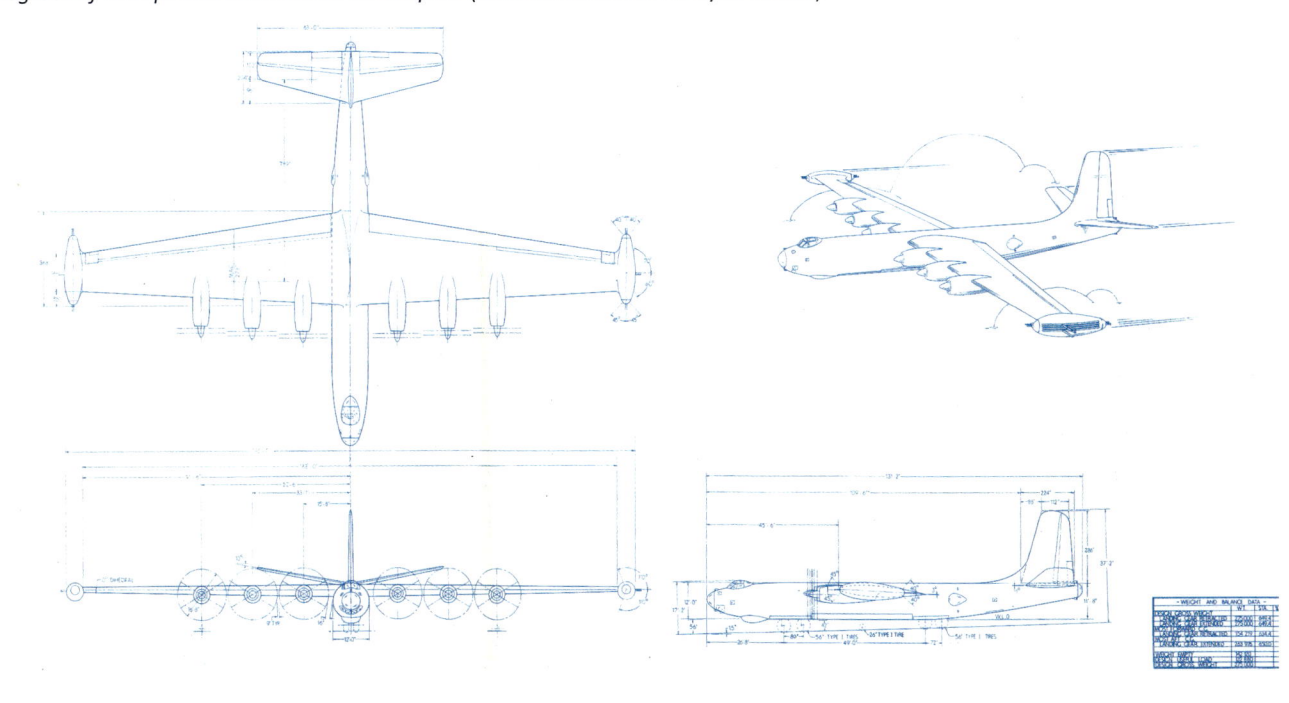

Model 462

The Convair and Martin proposals for the long-range heavy bomber competition in 1946 have already been discussed. Next comes Boeing's proposal, a design study for a heavy turboprop successor to the XB-36.

Boeing designed the Model 462 using the technology and experiences it had acquired from the B-29. Model 462 was, in fact, a very beefy B-29, approximating the size and weight of the XB-36 that was still a year away from first flight. The project underwent a series of model and design changes, political and budgetary twists and turns, complete program stops, cancellations, and restarts before eventually becoming the XB-52. This arduous birth all started with this concept of the Model 462.

The Model 462 proposal brochure of 27 June 1946, gives the stats on this giant bomber:

Wingspan	221ft (67m)
Length	161ft, 2 in (49m)
Design Gross Weight	360,000lb (163,293kg)
Power	6 Wright T35 driving 20ft (6m) diameter props
Bomb Load	44,000lb (19.958kg)
Crew	10

This is the second in a series of concept drawings of the Boeing Model 432. Later the Model 432 took on swept-back wings, and was re-designated Model 446. It is now starting to look more like the XB-47 that would eventually lead Boeing to the B-47. (National Archives via Ryan Crierie)

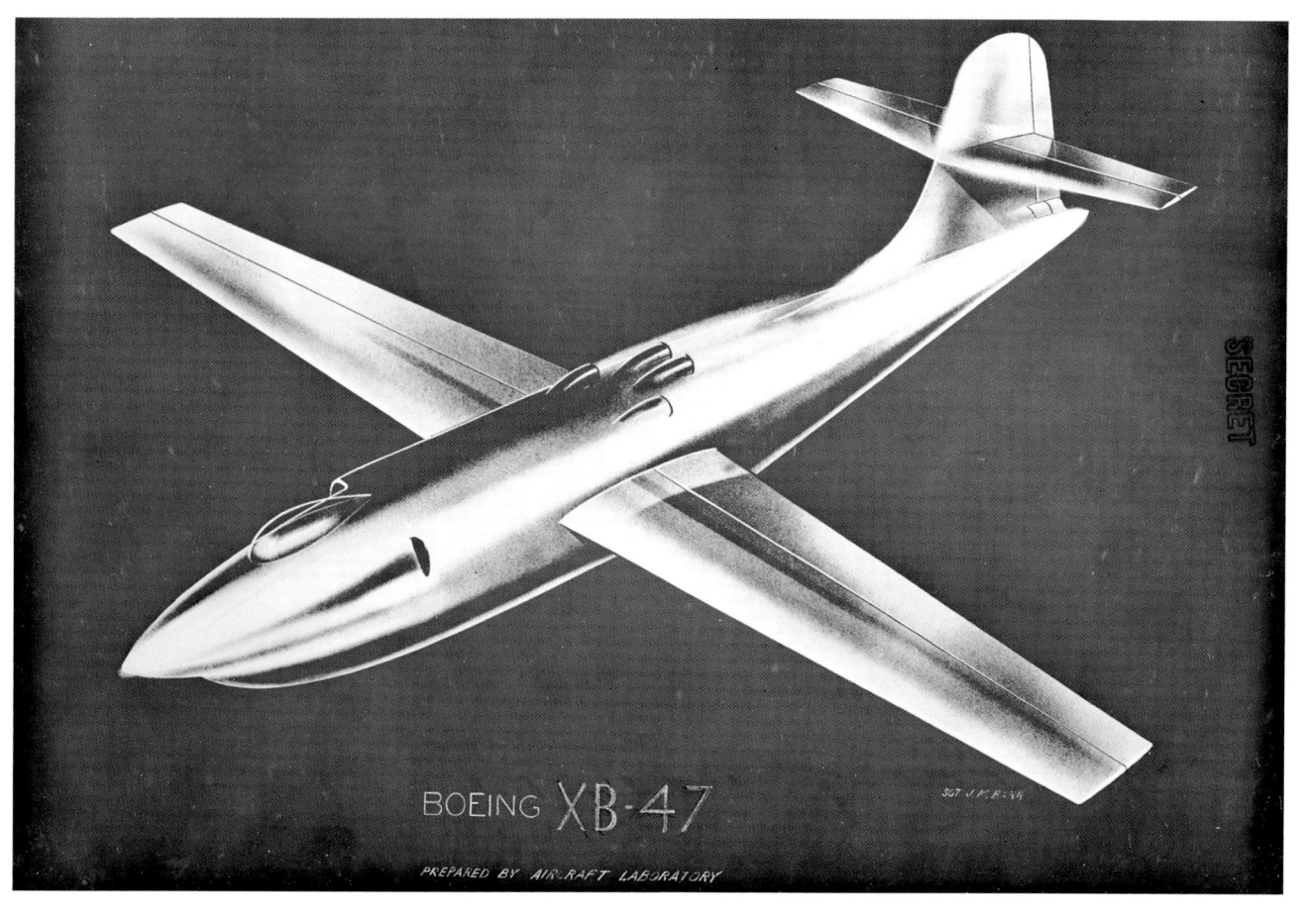

The Bombers Chapter 2

The in-house Boeing Model 462 model has the lines of a B-29 'on steroids'. It was derived from the experiences and technology garnered from building and flying the Superfortress. This iteration is the turboprop version of what would eventually be the XB-52. (National Archives via Jared A. Zichek)

A Boeing concept painting of the Model 462, which won the 1946 AAF (Army Air Forces) competition for a long-range heavy bomber. Although it was never built, it influenced the design of the XB-52. (National Archives via Jared A. Zichek)

America's First Jet Bomber

The North American B-45 Tornado was designed as a jet aircraft and made history as the first operational American jet bomber. Its humble birth showed the airframe of a propeller-driven aircraft outfitted with jet engines instead.

By 1943, the Army Air Force (AAF) had received intelligence of German aircraft powered by jet engines. The AAF asked the General Electric Company to start research and development on a more powerful engine than its prospective axial turboprop. This eventually resulted in the J35 and J47 turbojets. Then in 1944, the War Department requested aircraft manufacturers to submit proposals for various jet bombers, with gross weights ranging from 80,000lb (36,288kg) to more than 200,000lb (90,720kg). North American, Convair, Douglas and Martin answered the call.

By 1946 the USAAF, pressed for quicker results, decided to stop the contractor competition and review the designs. It was decided that the North American multi-jet-engine B-45 Tornado was to be put into production, making it the first American jet-powered bomber to enter production. It was also the first American jet bomber to reach operational service, and it was the first jet bomber to be refueled in mid air. And once the atomic bomb was made small enough, it was the first jet bomber to carry nuclear weapons.

The rush to put the B-45 in the air created structural changes that did increase its weight, which in turn gave it excessive takeoff distance. The first flight was 17 March 1947, and even so, an order for 193 was placed.

The improved B-45C model differed from its earlier variants by mounting the 1,200gal (4542.5 litre) fuel tanks on each wingtip. But the B-45 saw its best work as a reconnaissance aircraft, designated the RB-45C, and used in the Korean conflict. This aircraft was capable of flying highly classified photographic missions deep into enemy territory, and eventually was used to overfly many communist satellite countries and even the Soviet Union, itself – though because of Presidential non-approval these flights were operated by the Royal Air Force. The B-45 in essence, became the early predecessor of the Lockheed U-2 and SR-71.

Southwick, an American supplier of display models for the manufacturers made this RB-45C reconnaissance version for North American with a wood fuselage and fiberglass wings. The reconnaissance version no longer used the glass nose enclosure of the bomber. *(Allen Hess collection)*

The Bombers   Chapter 2

North American B-45 in-house wood model uses a high gloss lacquer paint to give this model great appeal. It is one of the earliest versions of the B-45 Tornado with its simple one-piece glass nose and canopy. This large model measures just over 4ft in wing span. *(Greg Barbiera collection)*

Another view of the same B-45 model showing the aircraft's highly conventional lines: straight wings, fighter-style canopy, and glass nose and tail gun position of a prop-type bomber. Note how the engine nacelles are faired into the wing. *(Greg Barbiera collection)*

The B-45 carried a crew of four: two pilots sitting in tandem, a bombardier, and a tail gunner. Only the pilots had ejection seats. In November 1948, the Tornado entered service with the newly created Strategic Air Command and remained in the Air Force inventory until its retirement in 1958. The B-45 was truly a pioneering airplane, serving proudly as America's first operational jet bomber until higher-performance aircraft could be built. Considering that the U.S. military relied on piston power for its long-range strategic bomber fleet through the end of World War II, it seems almost miraculous that the B-45 entered service only three short years after that War ended.

Nemoto, a Japanese supplier of display models for the manufacturers during the 50s and 60s, made this 1/72 solid wood version of the B-45C. Notice the segmented glass nose and wingtip fuel tanks typical of the C model. This early fuel tank carries stabilizer fins on the top front as well as the tail of the wingtip tank. *(Author)*

57

Chapter 2 — Secret US Proposals of the Cold War

Parasite Bombers and Fighters

The concept of protecting a slow mother ship aircraft by using nimble parasite fighters once it was over enemy territory first appeared in World War I. Both the Germans and the British experimented with fighters slung under dirigibles but, because of the aeronautical state-of-the-art at that time, the idea never seemed to pass beyond the experimental stage. In the 1930s, the Americans did the same with the Curtiss F9C Sparrowhawks carried by their giant lighter-than-air ships, and the Russians also tried carrying small fighters on some of their larger bombers.

In the United States the idea of using a parasite fighter continued after World War II with the McDonnell XF-85 Goblin, a small fighter that was to be carried in the bomb bay of the B-36. Actually, the intended mother ship was to have carried as many as three Goblins. These parasite fighters were small and rotund, with a fuselage just over 14ft (4.3m) long, with folding wings, with a span of only 21ft 1.5in (6m). Powered by a Westinghouse J34-WE-7 turbojet of 3,000lb (13.3kN) thrust, this fighter had a maximum speed of 650mph (1064km/h) and a maximum endurance of 1 hour and 20 minutes. Its combat ceiling was an impressive 46,750ft (14,249m) and it carried four .50-cal. machine guns. This was remarkable in an aircraft less than half the size of conventional fighters.

The US Air Force ordered two prototypes in March 1947, but by the time they were ready for testing, no B-36s were available. Initial flight trials started on 23 August 1948 using a modified B-29. Unfortunately, the small Goblin experienced severe buffeting from the bomber's prop wash while attempting to re-attach to the mother ship. On one test flight it crashed into its own mounting trapeze. The difficulty seemed insurmountable. If highly experienced test pilots could not cope with the task, there was no prospect of regular service pilots doing so, and the program was canceled. Yet the basic concept of parasite fighters still seemed attractive enough to continue exploring.

This in-house Douglas Model 1211 parasite bomber model is packed away in its carrying case. This is how the model was safely transported to meetings with military officials. The wooden box has provisions for holding the stand (not shown) and a smaller box for the parasite fighters. (John Aldaz collection)

The Bombers Chapter 2

The Model 1211 model is made entirely of metal so that it would not sag under the weight of the different parasites that would hang from the wings. To keep it sitting on its nose wheels, three large lead ballast plugs of lead were inserted into the nose. It is shown with Douglas XF4D and the Convair XF-92A fighters and a variant of Northrop's Snark missile with a low wing so that it could fit beneath the 1211. (John Aldaz collection)

This in-house model of the Douglas Model 1211 parasite bomber is set up with its mission specific pods. These pods could carry fuel, bombs, even photographic equipment manned by personnel. Note that this model came with contra-rotating turboprops and the ends of the nacelles had removable cone shape plugs, indicating that a option was available to offer additional four turbojets. (John Aldaz collection)

59

Chapter 2 | Secret US Proposals of the Cold War

By the early 1950s the Fighter Conveyor (FICON) program was initiated to test the feasibility of a larger parasite fighter. Tests were conducted using a modified Republic YF-84F Thunderstreak jet (previously designated as the YF-96) slung under an RB-36F-1 (re-designated GRB-36F). The bomber had a special trapeze mechanism built into its bomb bay. But now, instead of the parasite being used as an escort fighter, the mission was changed to a strike role and the parasite became a bomber-borne fighter-bomber. The new plan called for the heavy bomber to bring the faster, more maneuverable F-84 to within range of the target, then release the fighter-bomber that would deliver a nuclear bomb and then either return to the mother ship to hook-up, or fly back to its home base on its own.

By 1953, the parasite role had changed one more time from attack fighter-bomber to reconnaissance. The YF-96A-type Thunderstreak was replaced with the photo-reconnaissance RF-84F Thunderflash, modified and designated as the RF-84K for these tests. The speed and agility of the RF-84K could be utilized to overfly heavily defended targets, gather photographic intelligence, and report back to the loitering bomber just outside the range of enemy defenses. However, by 1956, the obsolescence of the B-36 coupled with the development of aerial refueling and the arrival of the high-altitude Lockheed U-2 spyplane helped bring the FICON program to an end.

In an effort to make the Model 1240 parasite bomber more attractive to the military, Douglas offered many more variations than its first offering of the Model 1211. Seen here are the Convair XF-92, Douglas XF4D, Republic RF-84, modified Bell Rascal missiles, various fuel tanks, both reconnaissance and cargo pods, and on its center section, another concept jet, the Model 1251-A. (Author)

An alternative parasite configuration of the Douglas Model 1240 shows a Republic RF-84 variant swung out under its trapeze in preparation for release. Similar to the FICON concept tested with the Convair B-36 and the Republic RF-84F. (Author)

Douglas Model 1211

During the XB-52 competition, Boeing and Convair had a head start as they were the proven, incumbent suppliers of heavy bombers to the Air Force. However, Douglas decided to push its luck and proposed another competitor for the XB-52, designated Model 1211. This was an ambitious offering, with at least 40 variations of the Model 1211 drawn up between 1949 to 1951.

This was an enormous intercontinental bomber aircraft. The first version was drawn up as a straight-winged bomber with four turboprop engines. Then, in following versions it appeared with swept wings and tails. Many of these iterations kept the turboprops, but later models were designed with six to eight J57 turbojets. The Model l211-J had four turboprops with counter-rotating propellers and optional four turbojets in the rear nacelles, It had swept wings at a span of more than 227ft (69.2m), a length of a little more than 160ft (48.8m), and was just under 45 ft tall (13.7m). It featured a variety of payload options including fighter/photo-reconnaissance parasite versions of the Convair XF-92A and the Douglas XF4D-1 Skyray. It could also carry modified Northrop XSSM-A-3 Snark missiles or massive fuel/cargo/personnel pods, one under each wing, for specific mission requirements.

The 1211 had a gross takeoff weight of 322,000lbs (146,057kg) and a range of 11,000nm (20,383km); well within the objectives of the USAF requirements. The speed was 450 to 500 knots (834 to 927km/h) and could reach a combat altitude of 55,000 feet (16,764m). All variations carried an 8,000lb (3629kg) bomb and the plane had outer under-wing fuel tanks with droppable landing gear, used to aid this large heavy beast on takeoff.

It was, however, turboprop powered and the Air Force had set its heart on jets. It is interesting to note that at the same time the Soviets were themselves working on a massive, swept-wing turboprop-powered bomber, the Tu-95 (NATO codename 'Bear') which is still in service over fifty years later.

Douglas Model 1240

Although the Air Force did not buy the Model 1211, it must have encouraged Douglas to continue the studies of the parasite bomber and apply it to a heavy transport version, for Douglas soon came out with another proposal, the twin-boom Model 1240 which used the same wing design of the 1211. In 1950 Douglas started development of the "All-Purpose Airplane" and by February of 1951 made a formal proposal to the USAF.

The Model 1211 had only one fuselage and an extremely long wing that was needed to carry such a heavy load. A double fuselage seemed to be a better design, simply to give the wing much more strength.

Multiple versions of the Model 1240 were offered which included the Model 1240-A, -B, -C, -D and the 1242. Depending upon its function the aircraft would vary in size with wingspans that ranged from 174ft (53m) to an astounding 347ft (105.8m) with a maximum gross weight of 977,000lb (443,167kg).

This three-view drawing is an early version of the Douglas Model 1240 parasite bomber. Notice that there are two canopies, one on each boom. The final configuration went to a single faired-in cockpit on the left boom as shown on the model. (Jared A. Zichek)

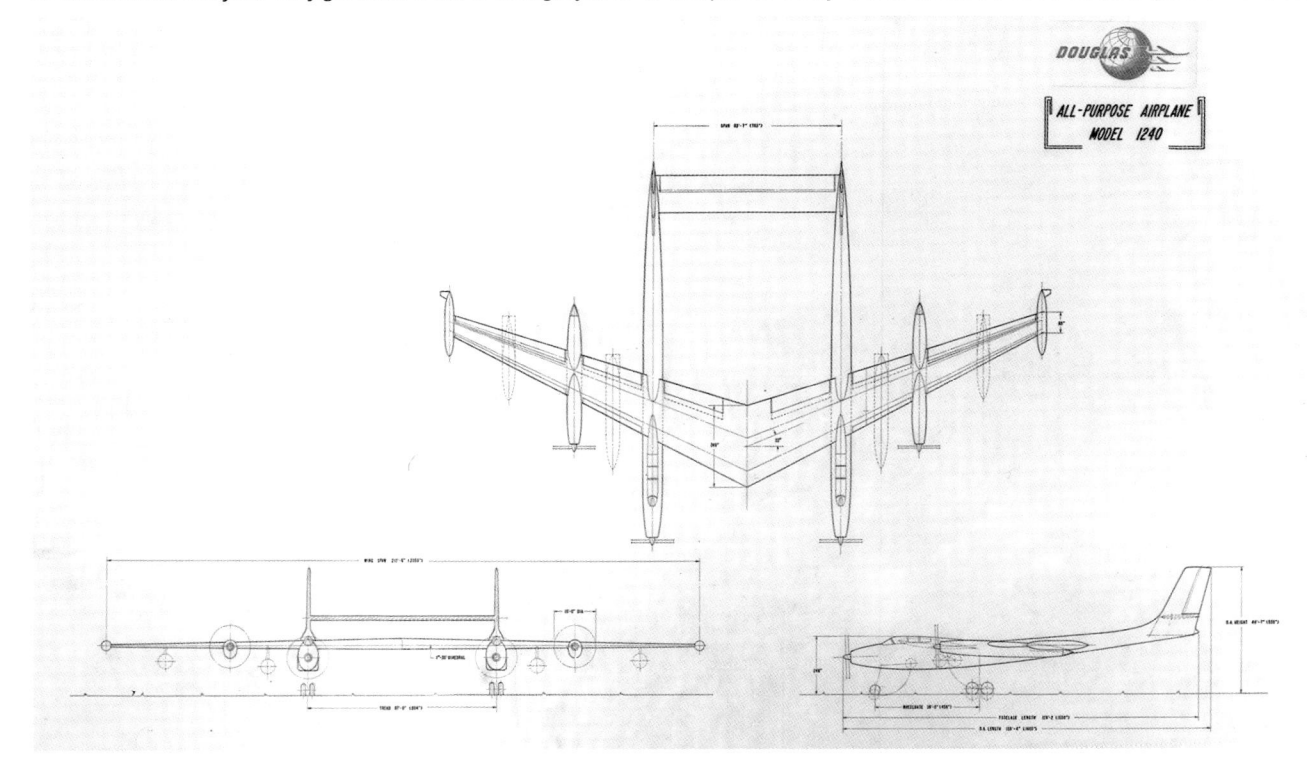

Early versions of the Model 1240 were offered with tandem cockpits in each boom. Eventually Douglas changed to a single side-by-side cockpit on the left fuselage only. The engineer drawings titled this newer version of the 1240 as the "XC Heavy Transport and Missile/Parasite Carrier". It was powered by four Pratt and Whitney XT34-P-10 turboprops with two optional General Electric 7E-XJ53-GE-X-25 turbojets in the rear nacelles.

Compared to the Model 1211, the Model 1240 could carry a larger variety of parasite fighters, reconnaissance and observation pods, missiles, fuel tanks, and assorted weapons under its wings and center section. Specifically it could carry parasite fighters such as the Douglas F4D-1 Skyray, Convair XF-92, Vought F7U Cutlass and the Douglas X-3. The X-3 was also offered as a pilotless missile along with the Northrop B-62 Snark, the North American Navajo and Bell Rascal missiles. Besides the fighters a Republic RF-84 Thunderflash for reconnaissance was added to the options.

The 1240 could be configured to carry very large pods in its center section that could house military equipment, personnel, a field hospital, a refueling station with flying boom, or a photographic reconnaissance unit complete with cameras and darkroom. There was even a pod with forward and rear firing air-to-air missiles.

Because of its dual fuselage configuration, the Model 1240 could carry even larger aircraft and one of the engineer drawings shows a Douglas A3D Skywarrior hung in the center section. But the Model 1240 also featured a very unique parasite attack bomber, the Model 1251-A.

This may have been the ultimate parasite Douglas described in its "XC Heavy Transport and Missile/Parasite Carrier" document, in the hope of swaying the military to buy not one but two of their proposals. Together the 1240 and the 1251A fulfilled the quest of being able to deliver the atomic bomb to the enemy, and return quickly and safely. The Model 1240 mother ship would carry the smaller parasite attack bomber to the enemy's border, safe from enemy detection, drop off the sleek Model 1251A, and then at supersonic speed the latter would deliver the atomic bomb to the target and head for home before any guns or interceptors could get a bead on it.

The Model 1251-A had a unique set-up in its configurations while heading to target and then outbound from target. Inbound the bomber was heavily laden with two turbojets suspended on pylons, two fuel tanks, two wingtip tanks, and the bomb. Outbound it would jettison everything and fly home in a clean configuration using its internal turbojet.

Of all the different configurations the Douglas parasite bomber could be made into, this had to be the ultimate presentation to the military, showing two concept models in one with an elaborate stand. Here the Model 1240 carries a second concept proposal, the Model 1251-A attack bomber. Once dropped by the mother ship, the supersonic Model 1251-A would deliver the deadliest of payloads and fly home on its own. (Author)

The Bombers Chapter 2

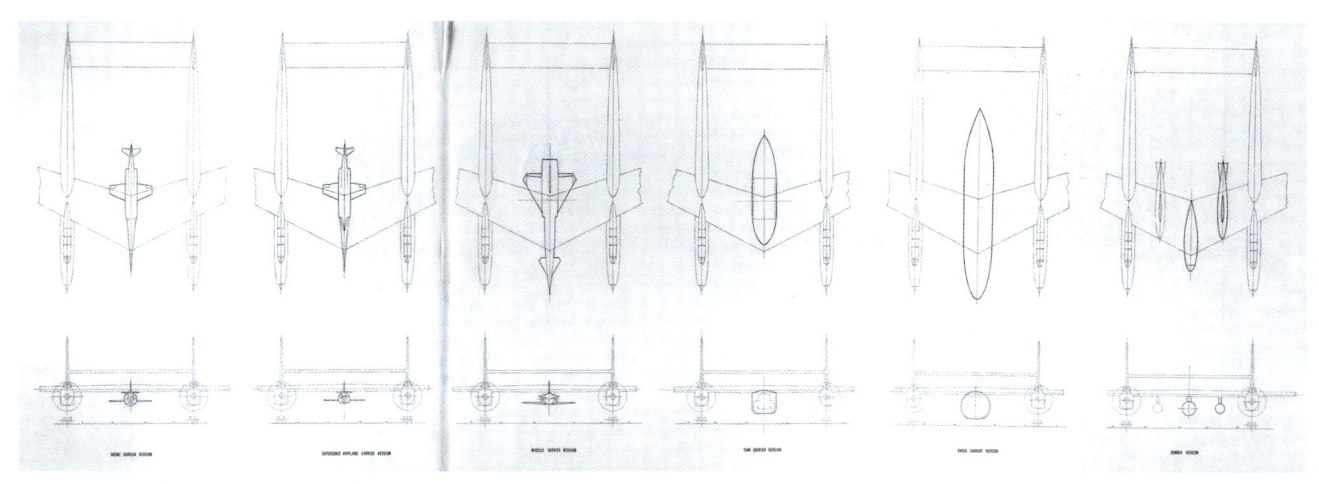

Engineering drawings depict different parasite configurations of the Douglas Model 1240. Showing just some of the many payloads that could be placed in the center section of the bomber, they include the X-3 drone, X-3 supersonic fighter, the Navaho missile, fuel tank and cargo/personnel carrier pods and even nuclear bombs. (Jared A. Zichek)

This model was given an interesting color-coding to its tip tanks, fuel tanks, jet pods and bomb, more than likely to be used during the sales talk for ease of explanation. The red colored wing tanks and atomic bomb of the Model 1251-A would be dropped on the inbound flight to and over target, while the blue under wing fuel tanks and, eventually, the blue turbojet pods would be released on the return flight, the aircraft switching over to its internal engine to complete the flight home. Landing gear was eliminated to save weight and the bomber would land on skids.

The Model 1251-A had a crew of three, a pilot, co-pilot/flight engineer, and a bombardier/navigator. It had a wingspan of 59ft 2in (18m) and an overall length of 90ft 10in (27.7m). The droppable jet pod engines and the internal engine were all General Electric 1-XJ53-GE-X25s.

These three-view drawings of the Model 1251-A depict the configurations of the attack bomber at the time of launch and on its cruise home. Originally the Model 1251-A had two skids for landing, this version shows a single skid, probably modified to save weight. Also shown are the cramped compartments for a crew of 3, pilot, co-pilot/flight engineer and bombardier/navigator, and the internal engine layout. (Jared A. Zichek)

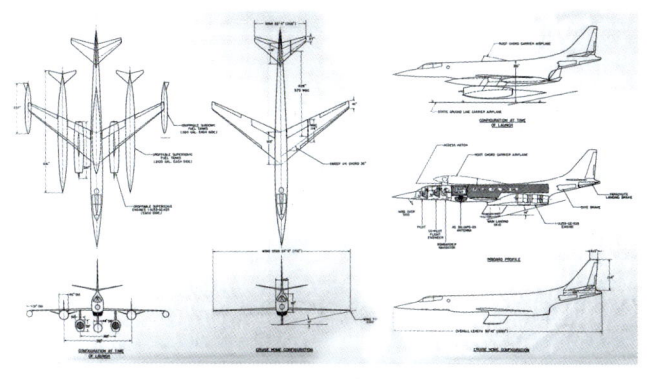

Below: *The Model 1251-A is shown detached from the Model 1240. The fairing that attaches the attack bomber to the mothership is molded onto the top of the attack bomber but in actuality would have remained on the Model 1240. (Author)*

Below right: *The stores under the Model 1251-A model are color-coded to indicate the order of release. The red wing tip fuel tanks and the nuclear bomb would be released inbound to the target. And the blue-colored under wing fuel tanks and eventually the engine pods would be released on the way back to base. (Author)*

The Douglas Model 1240 with its Model 1251-A parasite attack bomber made quite a pair in concept and presentation. But in the end they were turned down because General Curtis LeMay favored either Boeing's concept or the rival Convair XB-60 design. The latter was basically the company's B-36 fuselage married to swept wings and tail surfaces and powered by the same jet engines as the Boeing offering. Prototypes were ordered of both airplanes but in the fly-off competition, the Boeing aircraft proved markedly superior.

Atomic Bombers

According to an advertisement in an issue of a popular model airplane magazine, *Air Trails* for November 1951: "Locked in one pound of the fissionable material U-235 is energy to equal *6 million gallons of airplane gasoline!* It is no wonder engineers are planning to propel planes with atomic power…But an atomic power plant emits deadly radiation. The tons of shielding necessary for protecting men and instruments would be more than any existing plane could carry. A sky giant must be created… an airplane that will dwarf anything in the sky today."

This was the promise and the problem of nuclear power. But the thought that one pound of atomic material could replace millions of gallons of gas was too enticing for the US Air Force and the Navy. And so the lure of atomic power initiated the search for an aircraft powerplant that could utilize this abundant energy.

Interestingly the ad states the feelings and reflects the attitudes of the times when it says: "The aviation industry progresses rapidly. New designs, power-plants, operating procedures…every field of aviation demands men who have the schooling, creative ability, and the foresight that will enable them—not only to keep pace with aviation—but to lead it." These were the sentiments of a nation coming into the nuclear age while worried about another country possibly bent on world domination in the Cold War. Strategically speaking, the first nation to come up with a nuclear propulsion system would have a marked advantage over anyone else.

The idea of nuclear power for aircraft began with the start of both the atomic age and the jet age in the early 1940s. During World War II, Col. Donald Keirn, an Army Air Forces power-plant specialist, was assigned to work in the new field of turbine power. He consulted with Sir Frank Whittle, British jet engine pioneer, and became involved in completing the General Electric–built Whittle engine that powered America's first jet plane, the Bell XP-59.

With the development of the atomic bomb during the war, Col. Keirn began thinking of the possible application of atomic power in aircraft. Eventually, he climbed the ranks to Major General and was involved with the various agencies that were formed after the war to look over and develop nuclear propulsion.

In 1946, the Nuclear Energy for Propulsion of Aircraft (NEPA) was formed, and an Army Air Force letter of intent was sent to the participating aircraft engine companies to start the study and development of a nuclear propulsion system. The Fairchild Engine & Aircraft Company was selected to lead the investigations of feasibility and application of nuclear power to aircraft. It was also commissioned to develop components for this system. Among the companies under Fairchild, General Electric and Pratt & Whitney eventually became major contractors to develop nuclear-powered jet engines.

In 1951, General Electric started the Aircraft Nuclear Propulsion (ANP) project from its Aircraft Gas Turbine Division to investigate two possible power cycles, direct and indirect.

But the problems of powering aircraft with nuclear energy were many. First, little was known at the time of the effects of radiation on metals and humans. It was known that the reactor would require shielding to protect the crew. Second, the weight of a reactor at that time was significant. Technology had to be developed to make a smaller reactor. And finally, when the goal for all jet aircraft was to achieve supersonic speeds, the added weight of the shielding and the reactor penalized the function of speed to the point where a nuclear bomber would never be able to break the sound barrier. The only true advantage of having nuclear propulsion was to be able to fly for days on end without the need to refuel. Even so, this was the era of research, exploration and the possibility of new discoveries. Maybe a new way would be revealed that would make atomic-powered airplanes a reality.

Convair illustration of the NX-2 with the J57 turbojets under wing, indicating that this version employs the General Electric Direct Air Cycle nuclear engines. The J57s would be used for take-offs and landings and to fly over populated areas, cutting down on the radioactive contamination that the nuclear engines expelled during operations. (Scott Lowther)

The Bombers

Chapter 2

Once the Fairchild Company proposal was underway, the first step of developing a nuclear power plant was the Hardware Development Phase, which began in 1951. Simply put, the goal was to replace the combustible gases in the turboprop or turbojet with the super-heated air created by the atomic reactor. It was determined that there were two ways of providing power to these engines: One was to heat the air that goes through the turbines directly from the nuclear reactor, called the direct cycle, and the other was to indirectly heat air next to the reactor, the indirect cycle. General Electric was given the contract to develop the direct cycle power plant while Pratt & Whitney would develop the indirect cycle.

Development of the new engines would take some time, but by April 1955 the US Air Force issued the specification of Weapon System 125A which called for a high-performance nuclear-powered aircraft.

Convair NB-36H

Consolidated Vultee (Convair) was chosen to conduct preliminary tests using a modified B-36H as a test bed. Since this was the largest aircraft available with a bomb bay designed to hold 40,000lb (18,144kg) of bombs, it could easily hold the 35,000lb (15,876kg) reactor. Convair was to fly the reactor and gather data on the effects of air scatter, radiation patterns produced by flying aircraft, and observe the radioactive dosage to crew and ground personnel. Also to be observed were the effects that low and high altitudes would have on the reactor.

A B-36 with a damaged nose section became the subject aircraft for this project. That aircraft's forward fuselage was replaced with a new crew compartment fitted with the lead shielding needed to protect the crew. This 12-ton (10,886kg) enclosed compartment had no windows except for a small airliner-type windshield through which the pilots could see only see straight ahead. An 8,000lb (3629kg) lead disk was also inserted between the reactor and the crew compartment for added security. To help vent and cool the reactor, large air inlets were added to both sides of the fuselage.

This modified bomber was designated the NB-36H and made its first flight in September 1955. In 1957 it was retired after flying 47 missions. It is interesting to note that the reactor never provided power to the engines of the NB-36H. The aircraft flew solely on the power of its original Pratt & Whitney R-4360 piston power plants. Rather this was considered a dress rehearsal for observing what it was like to fly a large airplane with a nuclear reactor onboard, powering it up once the aircraft was high over unpopulated desert, bringing it back to base, and then dealing with the storage of a hot radioactive reactor on a flight line.

Convair's model of the NB-36H, a modified B-36 showing a redesigned nose to shield the crew compartment from radioactivity. It carried a 35,000 lb. reactor that hung in the bomb bay on a large hook so that it could be jettisoned quickly in case of an emergency. The nuclear reactor never powered the aircraft, it was carried aloft simply to test the reactor at high altitudes. Note the air intake vent modification at rear fuselage, just behind the inboard propellers, used to help cool the reactor. (George Cox)

Saunders-Roe Flying Boat: The Nuclear Princess

During this period the Consolidated Vultee Aircraft Company officially became Convair. And in 1956 and 1957, engineers started considering ways of applying a nuclear reactor/turboprop to an existing aircraft. One of the first proposals was to refit a large British flying boat, the Saunders-Roe Princess. The Princess, first flown in 1952, was the ultimate commercial flying boat design. At that time it was the heaviest all-metal passenger transport and the largest gas-turbine powered aircraft. However, by the time it flew the era of long-range passenger flying boats was at end, made obsolete by the development of land-based aircraft capable of trans-oceanic flights.

The Princess was a ten-engined aircraft (eight of which were coupled in pairs). Although slightly smaller than the B-36, it had a deeper fuselage, and with a wingspan of 220ft (67.1m) and a length of nearly 150ft (45.7m), it was one of very few aircraft able to accommodate a heavy nuclear reactor.

Plans were drawn up to make either a four-engine version or a six-engine version. These were to be a combination of more powerful turboprops and nuclear-driven engines. The plans in this book actually indicate both versions were considered. One half of the Princess shows three engines on the right wing representing the six-engine configuration while the left wing shows two engines from the four-engine configuration.

Chapter 2 Secret US Proposals of the Cold War

Convair's in-house model of the Saunders-Roe Princess flying boat nuclear proposal. A nuclear reactor carried within the fuselage would power only the two larger inboard turbine engines. This aircraft was considered by Convair because its size was large enough to hold the heavy nuclear reactors of the time. (George Cox)

Top view of the Saunders-Roe Princess wing shows the new configuration of its engines. Convair's in-house proposal model has the four smaller T34 chemical engines outboard with the larger T57 nuclear powered engines inboard. (George Cox)

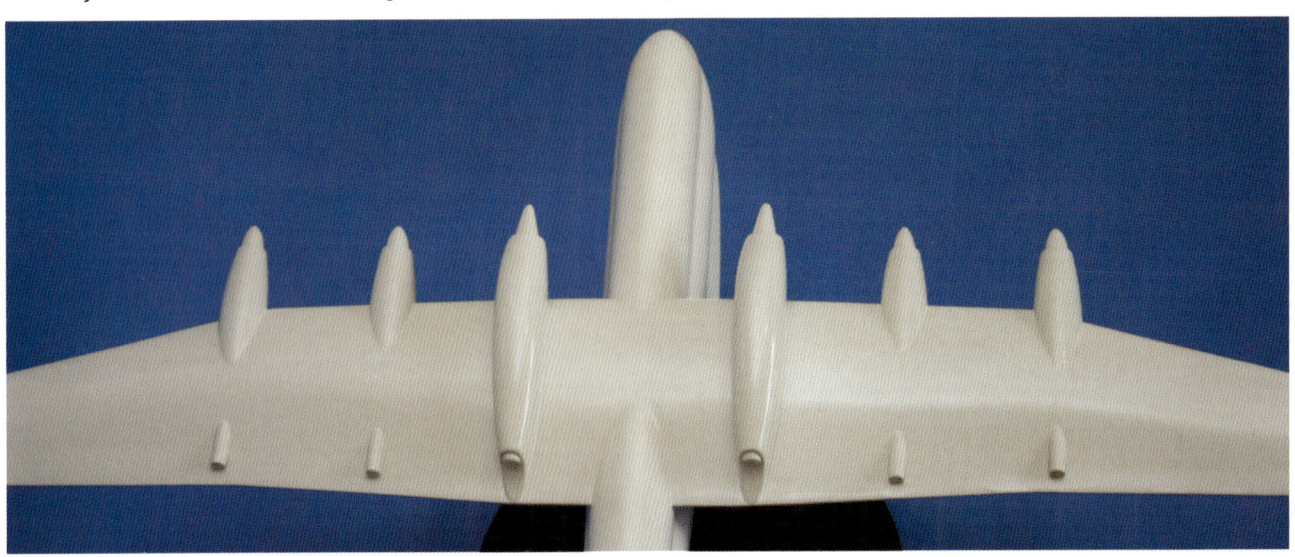

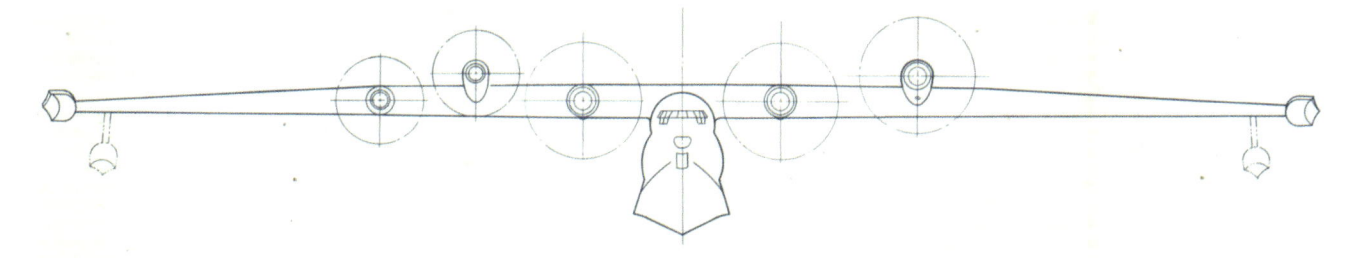

Convair front-view drawing showing two different engine configurations for the Saunders-Roe Princess flying boat. Left side (right wing) shows three of the six-engine version while the right side (left wing) would be for a four-engine configuration. Note the two smaller outboard engines used on for a six-engine configuration. (Scott Lowther)

The inboard layout of the Saunders-Roe Princess Flying Boat is shown in engineering drawings from Convair. The reactor installation within the fuselage of the Princess is of particular interest. (Scott Lowther)

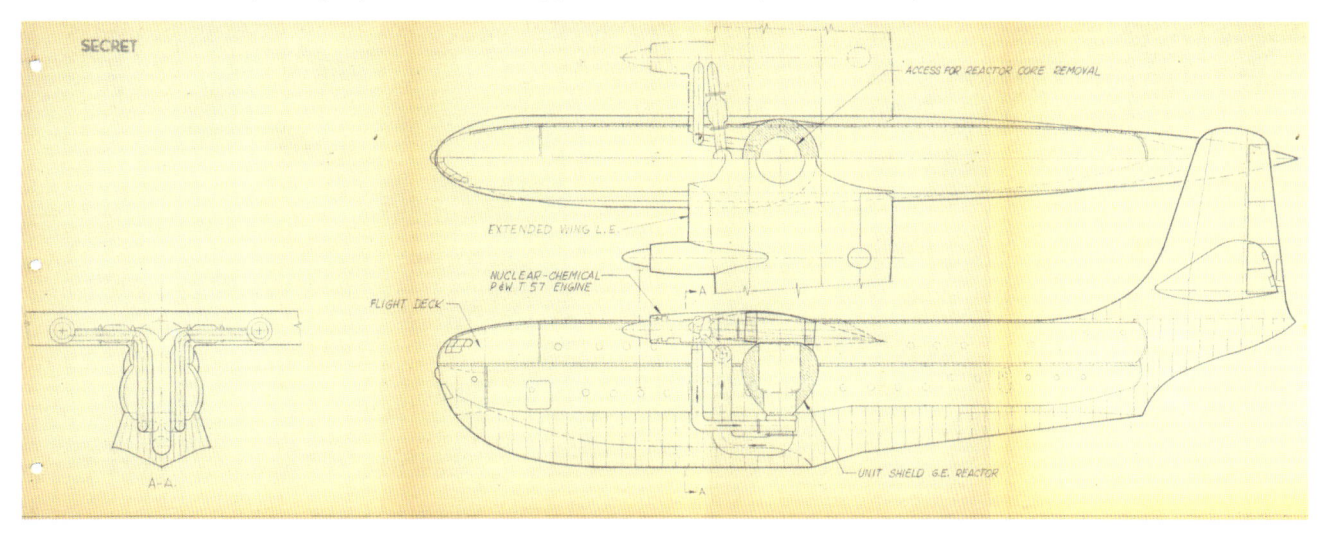

The six-engine variant was to have the four outboard, T34 turboprops conventionally-fuelled, while two inboard T57 turboprops were nuclear-powered. The four-engine version would be equipped with four T57 turboprops; the outboard ones would be conventionally-fuelled and the two inboard T57s to be nuclear-powered.

The proposal never went beyond the planning stages, but a model still exists of the six-engine version. It was found by a collector at a San Diego swap meet in 1998. Being a Brit he was familiar with the Saunders-Roe Princess, and initially thought he had acquired a model of the standard aircraft – something of a rarity in its own right. However, on closer inspection he realized he had never seen it with the engines configured in this particular way. After contacting a retired Saunders-Roe employee (who turned out to have been the project engineer for both the original Princess and its proposed nuclear variant) the collector discovered the model's true identity. The engineer who was greatly surprised to hear of the model's survival, explained that he had actually taken the very same item out to California to support a presentation to the US Navy some 40 years earlier!

Convair NX-2

From a press release by Convair in 1959: "First US nuclear-powered aircraft to be built for the Air Force by Convair Ft. Worth, will have a subsonic canard configuration and will be about the same size as the late model B-52s, weighing around 450,000lb (204,117kg). Model…can accommodate either General Electric Direct Air Cycle or Pratt & Whitney Indirect Air Cycle nuclear engines without major modifications. Present Air Force schedule calls for extensive flight tests using conventional engines with hydrocarbon fuels before the first flight on nuclear power late in 1965. This schedule is based on continuation of the current expenditure rate of about $150 million per year and no major, unforeseen technical difficulties."

According to the press release, the NX-2 could work with either the General Electric direct-cycle GE X211 engines or the Pratt & Whitney indirect-cycle engines. Both engine types had dual capabilities, either working as conventionally powered jets or on nuclear power. The NX-2 would take-off using conventional power and, on reaching altitude, the reactor was fired up and the engines would then fly on nuclear propulsion. Thus the conventionally-fuelled jet engines would be used only for take-off, the climb to altitude and landing.

Chapter 2 | Secret US Proposals of the Cold War

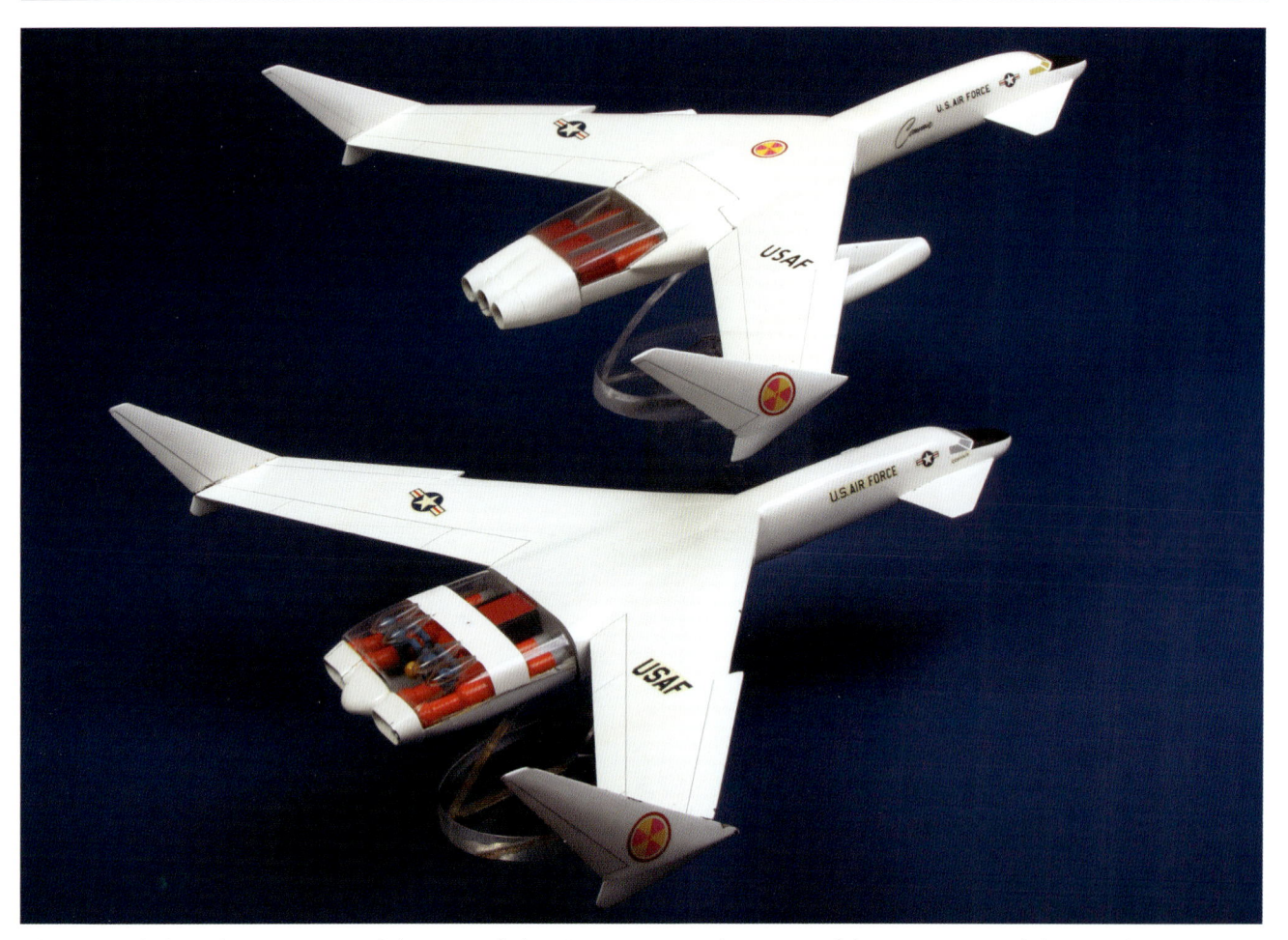

Above: *A side-by-side comparison of the General Electric Direct Air Cycle version of the Convair NX-2 (above top) and the Pratt & Whitney Indirect Air Cycle version (below bottom) shows that both of these in-house models exhibit elegant design and form. (Jonathan Rigutto collection)*

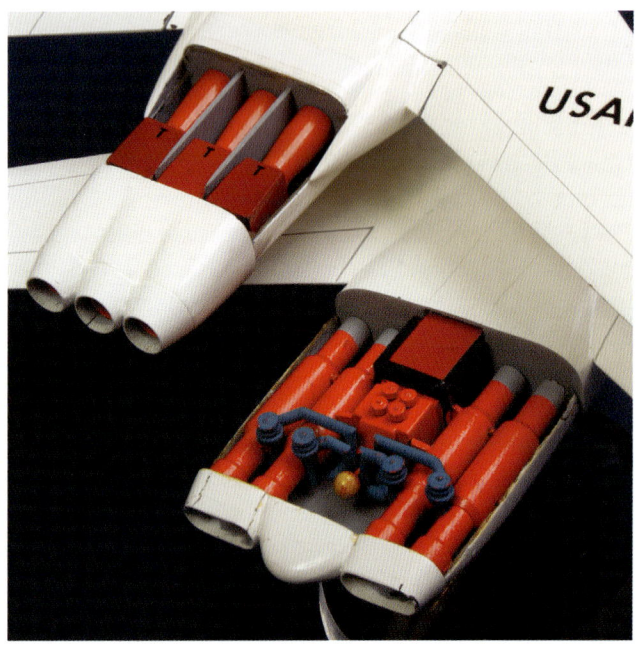

Comparison of the Convair in-house wood models of the NX-2 demonstrating the two different types of nuclear engines. The top model shows the General Electric Direct Air Cycle and the bottom one exhibits the Pratt & Whitney Indirect Air Cycle. (Jonathan Rigutto collection)

The canard configuration was a key feature of the Convair NX-2. The designers wanted to keep the crew compartment as far away from the reactor as possible. Their solution was to stretch out the forward fuselage and replace the elevators with the canards, as shown on these Convair-built wood models. (Jonathan Rigutto collection)

68

The Bombers

Packed in its shipping box is the Convair NX-2 in-house wood model that converts from the General Electric Direct Air Cycle version (shown on the model) to the Pratt & Whitney Indirect Air Cycle version. Notice the two J57s strapped in the upper part of box and the Pratt & Whitney replacement engine in the bottom corner. (Jonathan Rigutto collection)

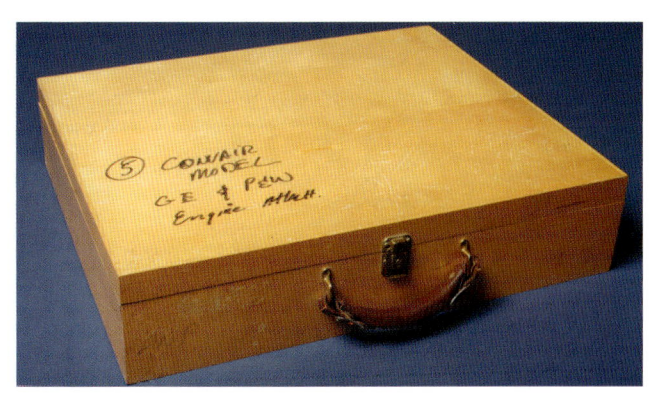

A Convair NX-2 shipping box holds the model that can display the General Electric or Pratt & Whitey nuclear engines. After the NX-2 program was cancelled, the two boxes containing NX-2 models were found in the trash bin and salvaged by one of the engineers who worked on the program. (Jonathan Rigutto collection)

The GE direct-cycle power plant was a "dirty engine." After the three jet turbines had been supplied hot air directly from the reactor, they would spew out radioactive contamination through the exhaust. As a measure to help reduce contamination, and for performance insurance as well, two under-wing J57s were used to help on take-off, climb-to-altitude, and landings. On the other hand, the four Pratt & Whitney engines did not pollute and so there was no need for the extra J57s. So the model without the under-wing J57s indicates the Pratt & Whitney indirect-cycle engines. Convair went on to make nuclear proposals of some of its jet seaplanes as well.

Northrop "BETA 1"

In the early 1950s Northrop studied some possibilities for its flying-wing design as a nuclear aircraft. It placed the crew compartment on one wingtip in an attempt to keep the crew as far away from the reactor as possible. Northrop also did a study of a nuclear parasite bomber. A slightly different version of this was offered by the Hawk Model Company as a "BETA 1" bomber (see the sidebar "Two Models that Reflect the Times").

Over the span of about 15 years, the journey of trying to make atomic propulsion work proved to be a difficult and frustrating road. The number of redirects and reorganizations of the program, the starts and stops of funding, and the political in-fighting among the military and the politicians ultimately contributed to its slow grinding halt. But not before one final success: in December 1959 the GE HTRE-3 reactor connected to a jet engine performed according to predicted specifications, albeit in the workshop. But by 1960–1961 advances in other options passed up the idea of having a nuclear-powered aircraft. The development of missile-launching nuclear submarines, the ability of air-to-air re-fueling, the proliferation of strategic airbases surrounding the Soviet Union, and ultimately, the ICBM, all eroded the case for the nuclear bomber. Moreover, the consequences of a crash in a populated area was a concern that could never be overcome. The Kennedy Administration put an end to all funding and closed the program in May 1961.

Once a major program ends, at times there seems to be no regard for all the hard work that had been done. It may be a case of sour grapes, or maybe there were orders from the government to destroy all classified materials. But in the case of two highly-detailed NX-2 models, they were summarily relegated to the dumpster. Fortunately they were discovered by a Convair engineer, who had worked on this program, while walking to the parking lot on his way home. He rescued them from the dumpster, but was then stopped by the gate guards and not allowed to leave the plant with them for they were still considered classified material. The engineer went back and put them in his office until a year or two later when the program was finally declassified, then took them home.

Below left: *Northrop illustrations show the different studies/configuration of its nuclear version of the flying wing. The two middle illustrations show the crew quarters, hung out on the wing tips as a means of keeping the greatest distance between the crew and the reactor.* (Scott Lowther)

Below right: *Northrop offered another idea for the use of nuclear power by creating this illustration to showcase a tanker version of its nuclear bomber design.* (Scott Lowther)

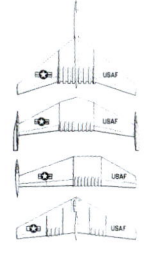

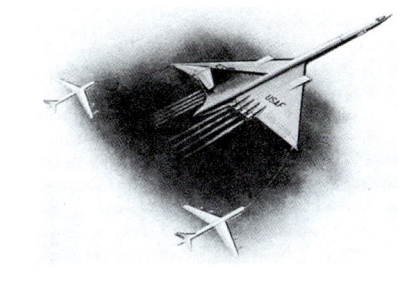

A full page advertisement from the November 1951 AIR TRAILS, a popular airplane hobby magazine, entices students to join the St. Louis University's technical college. (Author)

The Bombers   Chapter 2

Models Reflect the Times

The model airplane industry historically has used current news-making events to help sell their product; Charles Lindbergh's Spirit of St. Louis *solo crossing of the Atlantic; Wiley Post's* Winnie Mae *circling the globe in eight days; any race plane breaking world speed records; all these happenings were exploited to promote and sell new lines of model airplanes. Cold War headlines were certainly no exception.*

Russian Nuclear Bomber by Aurora

During the Cold War the Soviet Union and the United States had entered into technological game of one-upmanship. But, much of what was happening in the Soviet Union was unknown. With its vast landmass, a culture of secrecy and lack of public information, developments could take place far from prying eyes and the West could only speculate about what was going on However, the Russians would shock the West by being the first to deploy a combat ready, high-performance, jet fighter, the MiG-15, in the Korean conflict. Nonetheless, the United States citizens continued to believed they were the world leader in all modern technologies during the 1950s. After all, they had every comfort of life: television, washing machines, cars with giant tailfins, even Disneyland!

Americans looked upon the Soviets as a nation that struggled to be modern but, in fact, still lived in the times of the Czar. That image rudely disappeared when all the world awoke one day to the transmitted beeping of the first satellite circling the earth, a satellite launched by the Russians in October 1957 called Sputnik.

The following year, Aviation Week *magazine, a highly regarded trade publication, came out with the next bombshell. In their December 1958 issue, in an article entitled "Soviets Flight Testing Nuclear Bomber," Aviation Week claimed that the Soviets now had the world's first nuclear bomber: "A nuclear-powered bomber is being flight tested in the Soviet Union. Completed about six months ago, this aircraft has been flying in the Moscow area for at least two months. It has been observed both in flight and on the ground by a wide variety of foreign observers from Communist and non-Communist countries…The Soviet aircraft is a prototype of a design to perform a military mission as a continuous airborne alert warning system and missile-launching platform." The article was replete with pictures and a three-view drawing of the bomber.*

The Aurora plastic model of the Russian Nuclear Powered Bomber is based on the three-view illustration provided by *Aviation Week*. As a nuclear aircraft, it follows the tradition of employing an extended fuselage as a means of distancing the crew compartment from the nuclear reactor. *(Allen B. Ury)*

Aurora quickly released a model of the Russian nuclear bomber in order to take advantage of public awareness and a tremendous PR opportunity. This followed the 1957 announcement that Russia had succeeded (where the U.S. had failed) in building the first flying nuclear-powered aircraft. *(Craig Kodera)*

Chapter 2   Secret US Proposals of the Cold War

With Sputnik orbiting the Earth and now the first Russian nuclear bomber beating out the Americans, the Eisenhower Administration was feeling the pressure to accelerate the funding to the Aircraft Nuclear Power agency.

Jumping into the fray with both feet, the head of the Aircraft Nuclear Propulsion Office, USAF Maj. Gen. Donald J. Keirn stood before a congressional hearing and in stern words painted this threatening scenario, "Imagine … a fleet of enemy high-speed aircraft continuously patrolling the airspace just outside the early-warning net, capable of air–launching a devastating missile attack, followed by high-speed penetration attacks on our hardened installations." Our modern American world was now turned on end.

Having had our technological rug pulled out from under us, what better time for the American hobby industry to capitalize on this astounding news and produce a model of the now-infamous Russian Nuclear Bomber. And that's exactly what one of the leading plastic model manufacturers did. In 1959 Aurora Plastic Models of West Hempstead, New York, quickly issued its version of this bomber based on the pictures and three-views published by Aviation Week. It was a simple model with few parts and is a good reproduction of the three-view. The scale is unknown but was somewhere near 1/144.

Today we know that the Russian Nuclear Bomber never existed. In fact, it was a medium-range strategic bomber, the Myasishchev M-50 (the 'Bounder' as designated by NATO). At the time it looked very futuristic and menacing, but it is now known that its performance fell well short of Soviet expectations and it never entered production. The article and accompanying critical editorial in the 1958 Aviation Week was actually self-serving intelligence created to upset the military and pressure the politicians to keep the ANP well-funded and alive. It worked for a little while.

"Beta 1" Nuclear Bomber by Hawk

A competing plastic model company also took advantage of the national angst and produced its own atomic bomber, in 1959. But to balance the situation and make sure that the Americans did not feel left out, the Hawk Model Company produced the "Beta 1" Atomic Powered Bomber XAB-1. It was a model loosely based on the Convair XAB-1, which also has some elements of Northrop's proposed Nuclear Bomber.

*But just to make sure that it was not inadvertently giving away any military secrets, Hawk posted the following notice in the instruction sheets: "Incorporated in the design of the Beta 1 are the engineering criteria established for an atomic airplane. The design, although hypothetical, is completely within the realm of possibility. In no way does Hawk Model Company intend to imply that such an airplane exists or is on the drawing boards. However, what is presented here has been checked by one of the United States' leading aircraft companies and has been declared entirely sound and possible. Therefore, this airplane **could** exist."*

The model features retractable landing gear, parasite fighters that could detach from the tail, and clear red plastic "flames" that came out of the atomic engines. The instruction sheet also contains a diagram of a simplified, but essentially workable nuclear engine. The scale of this model is 1/188.

By the late 1950s, the concept of nuclear power represented the epitome of futurism. Nuclear reactors were now aboard submarines and would soon power super-sized ships including the world's largest aircraft carriers. Land-based nuclear powerplants would someday supply electricity to America's cities. For aircraft, however, nuclear propulsion was destined to remain an unfulfilled dream.

Hawk Model Company based its plastic kit of the "Beta"-1 Nuclear Bomber loosely upon the Convair XAB-I nuclear bomber concept with the added feature of parasite fighters. This unusual subject had a limited run which now makes it a rare and highly collectable model. This is a replica of the Hawk model made by Fantastic Plastic. *(Allen B. Ury)*

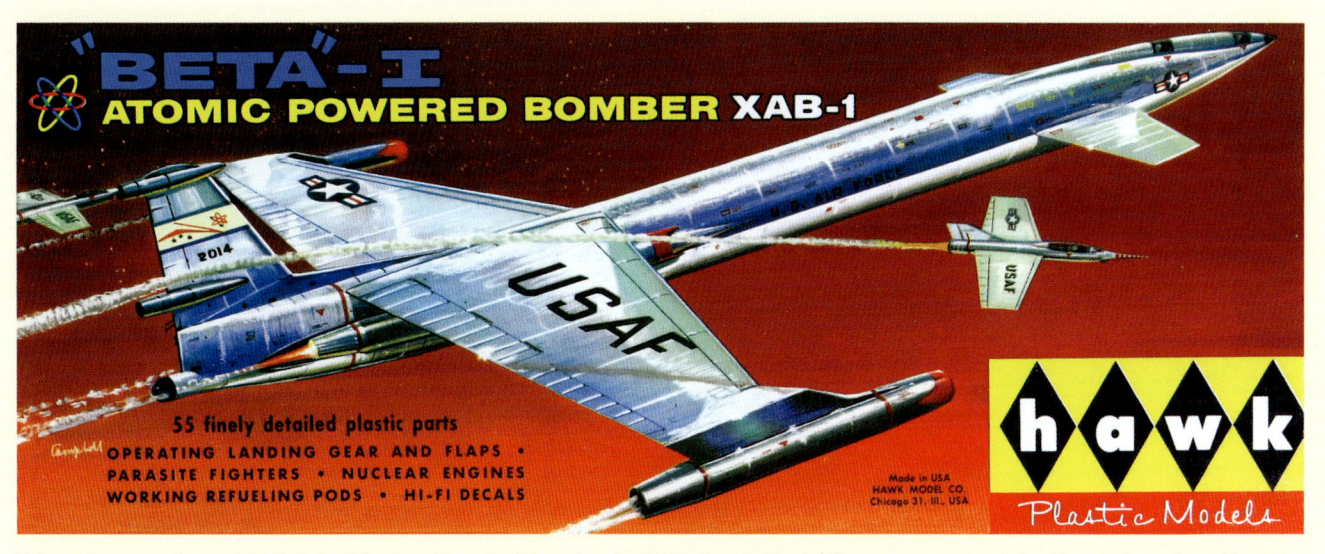

When Aurora released its Russian Nuclear Powered Bomber in 1958, Hawk Model Company answered with this American Atomic Powered Bomber in 1959. Even better than the Russian aircraft was the fact that this bomber included parasite fighters to defend the mother ship. *(Craig Kodera)*

Supersonic Bombers

On 14 October 1947, Air Force Capt. Charles E. "Chuck" Yeager flew the Bell X-1 rocketplane to a speed of Mach 1.06 to become the first man to break the sound barrier. On 20 November 1953, NACA test pilot A. Scott Crossfield became the first man to fly Mach 2 in the Douglas D-558-II Skyrocket. The natural conclusion of the Air Force was to think that all aircraft could and should fly at these speeds, if not higher. The first calls in a mad rush for giant fleets of supersonic fighters and bombers soon followed, and thus the supersonic age began.

But creating and flying a large craft at those speeds was still a challenging business. Aerodynamic drag rises steeply as the speed of sound is approached, and with it comes the need for more propulsive force. Even fighters with the most powerful jet engines need to resort to afterburning or re-heat to achieve supersonic speed. This involves burning vast quantities of extra fuel in the exhaust downstream of the engine. It means that most aircraft can only sustain such a speed for a matter of a few minutes. That may not be a problem for fighters which only need a short burst of speed to reach their foe but it is useless for a bomber.

This constraint dictated that a large bomber would have to cruise to its target at subsonic speeds and only once near the target, "dash" at supersonic speed for both the bomb run and the immediate exit from the target area. The alternative would be for a subsonic bomber to carry a smaller craft, loaded with its nuclear weapon, and release this close enough to reach the target at supersonic speeds.

Boeing MX-1712

Many of the manufacturers, including Convair, Douglas and Boeing, envisioned these possibilities and submitted interesting proposals. In 1952 Boeing offered a proposal brochure for the MX-1712, a medium size bomber with multiple capabilities, that the Air Force could use as a supersonic bomber, a missile carrier, a reconnaissance or electronic countermeasures (ECM) aircraft.

The technology was simply not yet capable of providing a supersonic bomber with the efficiency to fly from the United States to distant targets without refueling, and at this time, air-to-air refueling was in its infancy. So Boeing was offering a medium bomber that could stage from advanced bases and, with quick and convenient fieldwork, be converted into any of the four roles it was meant to play.

Again, what is interesting to note here is how the art and the copy of the brochure played on the needs of the Air Force for this multi-tasking bomber, which could achieve total destruction with a high level of technology (even though some of that technology is not exactly clear). Upon opening the proposal brochure the first image that greets you claims that Boeing can provide the aircraft that delivers total destruction of strategic targets, "an achievement made possible by the advancements into the realm of supersonic flight."

The MX-1712 promises a supersonic weapon designed for strategic medium-range reconnaissance and destruction of enemy targets with maximum protection from detection and interception. It has a maximum speed of 1,060 knots (1964km/h) or Mach 1.85, "which

Chapter 2 — Secret US Proposals of the Cold War

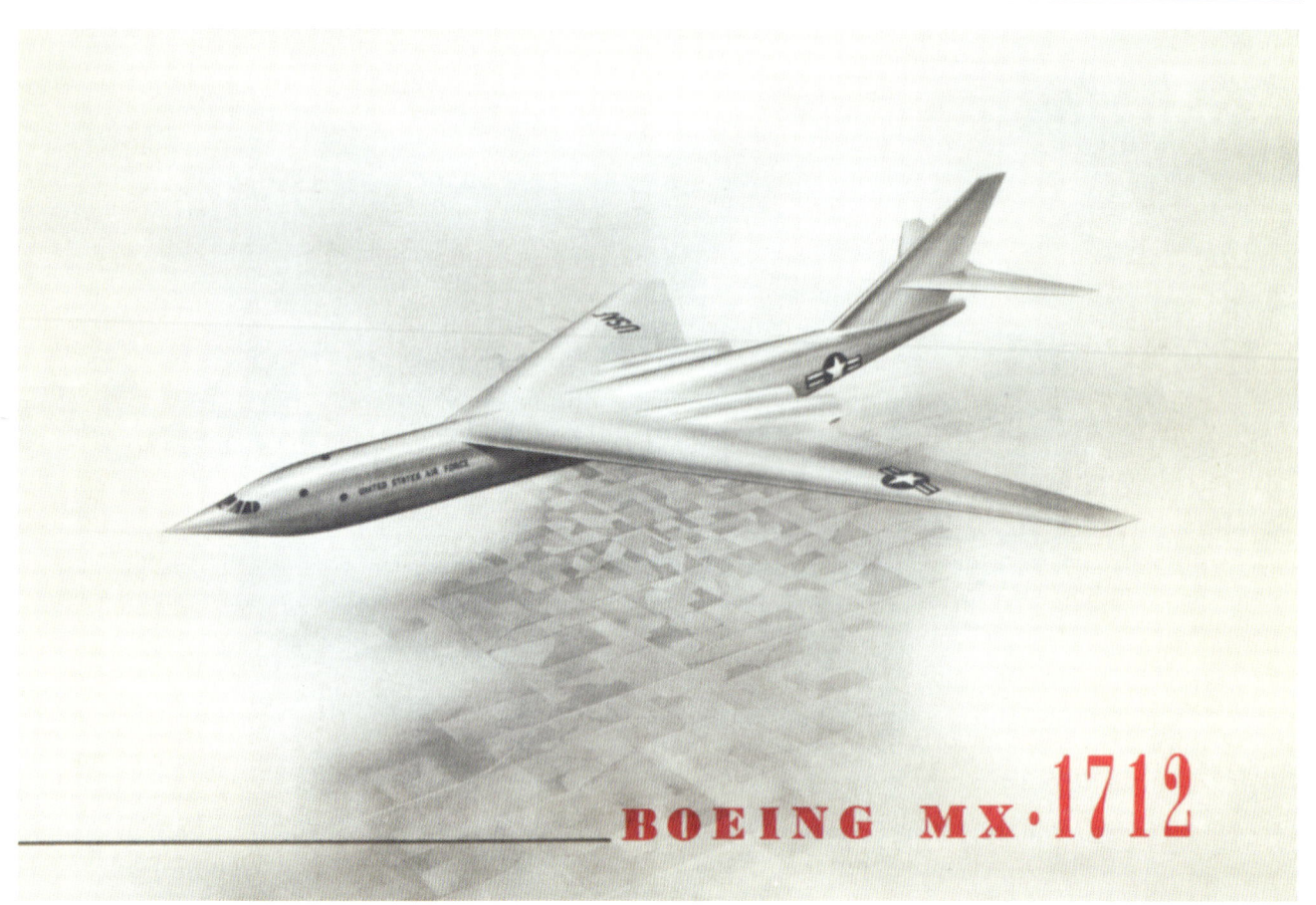

Above: *The cover page from the Boeing MX-1712 proposal brochure showed this nicely rendered illustration of the MX-1712 in flight.* (Scott Lowther)

Right: *A page from the Boeing MX-1712's proposal brochure diagramming two low altitude bombing techniques using a loop maneuver to deliver a "Special Weapon". Note the wording on the "loop" delivery tactic of the bomb states that it: "Will require the development of bomb aiming equipment."* (Scott Lowther)

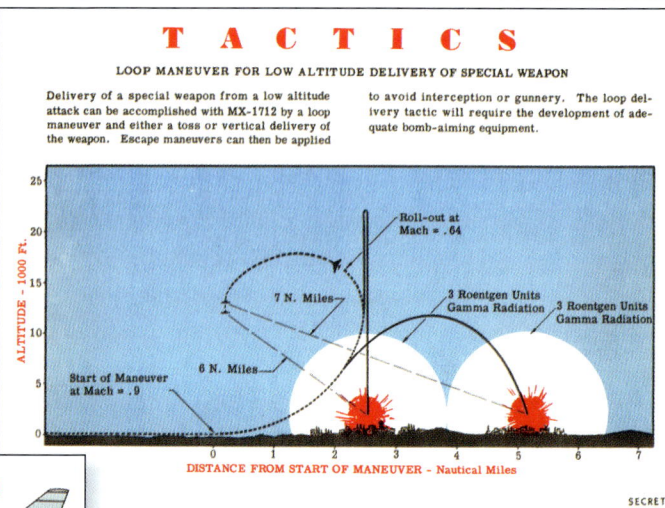

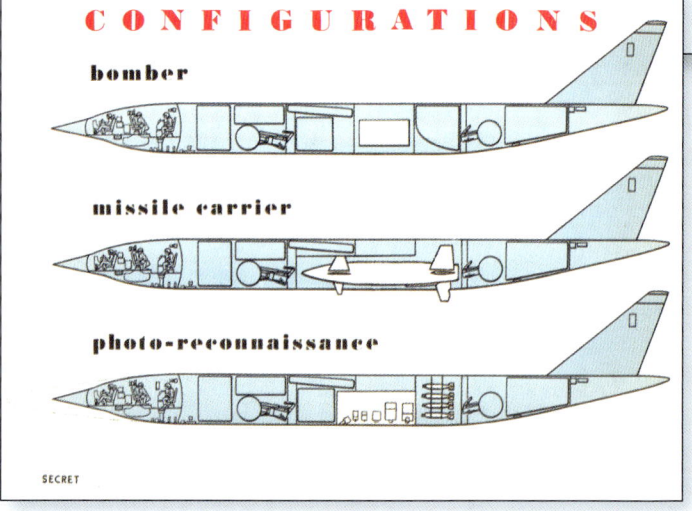

impairs the ability of defensive systems to detect, track, and control interception." It also touts an active and passive system of defense. The graphics are clean and simple, getting right to the point of what the MX-1712 was supposed to do.

Left: *Cutaway views from Boeing's MX-1712 proposal brochure show the three different configurations of the MX-1712 as a bomber, missile carrier and photo-reconnaissance platform.* (Scott Lowther)

The Bombers

Chapter 2

Convair MX-1964

On 1 February 1952, fully supersonic bomber requirements were issued under General Operational Requirements (GOR) SAB-51 (SAB stands for Supersonic Aircraft, Bomber). The Wright Air Development Center defined the full objective of a supersonic strategic bomber as being able to carry a 10,000lb (4536kg) load in 7 May 1952. The winning contractor was required to produce a bomber that had to fly most of the mission at supersonic speed, be able to carry the TX-13 "special weapon" (and other various atomic stores), and have an automatic bombing, navigation, and missile-guidance system that was both reliable and highly accurate.

During the early and mid 1950s Convair came up with the MX-1964 design concept, which evolved into the Mach 2 B-58 Hustler. But the Hustler could only go one way with its fuel load if it ever had to deliver the bomb. Moreover, Soviet defense capability was now able to intercept an aircraft flying at this speed. The Air Force wanted something bigger with global capabilities. The next step therefore was a bomber big enough to fly anywhere in the world that could achieve Mach 3. WS-110A was issued calling for a new bomber with chemically-powered engines that could reach the higher Mach number. Eventually this led to the amazing aircraft known as the XB-70.

If a subsonic jet bomber needed a lot of fuel, imagine how much more a supersonic one had to have to reach its target and return. Fuel capacity was still the problem so a number of manufacturers came out with proposals that had interesting solutions to large fuel requirements. One of the solutions to extend the range of the bomber was the floating wingtip. These were extensions attached to the bomber wing that carried additional fuel tanks.

Boeing Model 724

By 1955, Boeing already had proposed a number of supersonic bombers, and with the Model 724 it explored the different configurations of carrying its extra fuel. The Model 724-16 came out in 1957. It featured a conventional bomber powered with four General Electric X279A engines and floating wingtip tanks attached to each wing. The whole affair looked like three aircraft flying in formation.

Boeing's in-house Model 724-16 model with the floating-wingtip tanks detached. This bomber concept, which came out in 1957, is powered by four General Electric X279A engines. (Boeing via Tony Panopalis)

Boeing built a very impressive display of the Model 724-16 with two floating-wingtip tanks, which extended the range of the bomber. Although the floating-wingtip tanks look like conventional aircraft, they are actually unmanned flying fuel tanks. (Boeing via Tony Panopalis)

Chapter 2　　　　　　　　　　　　　　　　　　　　　　　　　　　　Secret US Proposals of the Cold War

North American NA-239

The North American NA-239 was the start of the design exploration for the XB-70, and much has been written about this amazing bird. The XB-70, however, went through several iterations before its final configuration was settled.

In 1955, the NA-239 was offered with the same idea of floating wingtips to carry the extra fuel, but different from the Boeing application. The floating wingtips had a span of 49ft (14.9m) and a length of 92ft (28m) each. Imagine how cumbersome this bomber would have been going down the runway with these floating tip tanks attached to each wings (not to mention how much wider the runway would have to be).

This in-house wood model of the North American NA-239 was an impressive representation of the bomber with its floating wingtips for extended missions. These extra wing panels would have created huge problems for ground handling. (Author)

An interesting view of the underside of the North American NA-239 with its floating-wingtip tanks for extended missions. When the North American photographer took this picture, the upright stem of the stand remained in the photo while the rest of it was cut away. Note the missiles under each of the large engine air intakes. (Author)

76

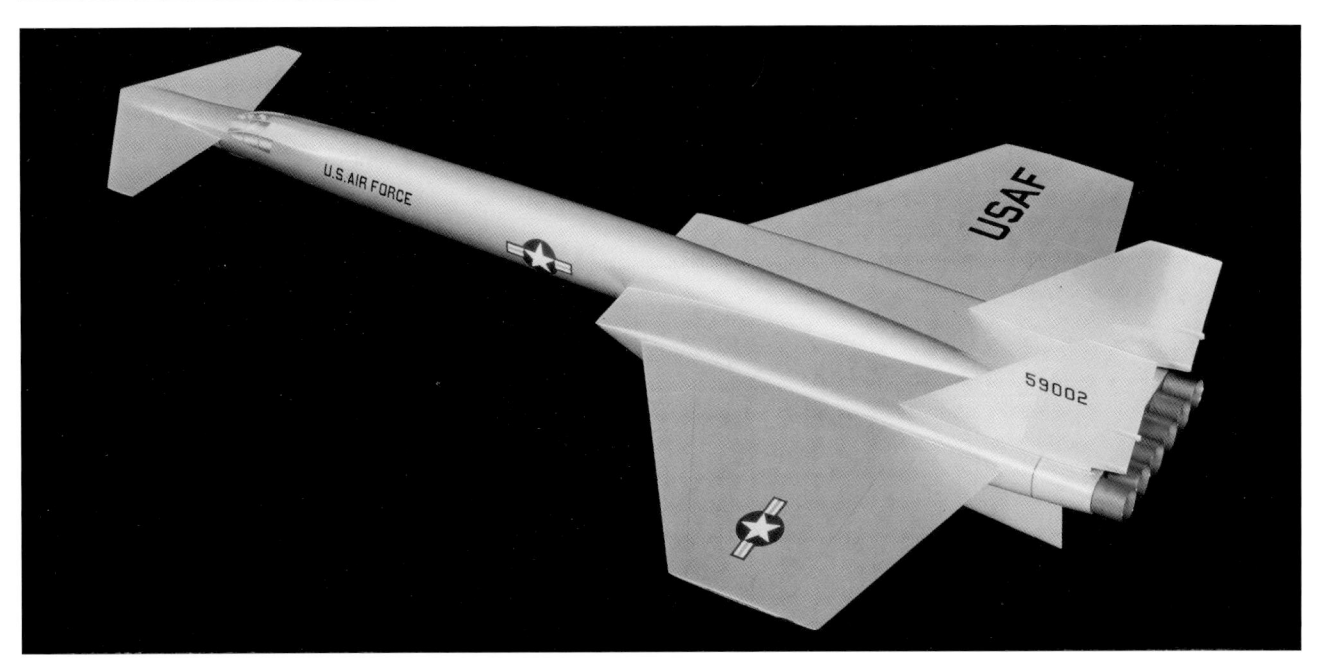

The NA-239 model is shown without the floating-wingtip tanks. This wood model, built by North American, was one of the first concepts for the XB-70. Note the six engines on this model. (Author)

North American XB-70

The first XB-70 Valkyrie made its maiden flight on 21 September 1964 and the second aircraft flew on 17 July 1965. However, its fate was already sealed. Despite giving every indication of meeting its promised design performance – and demonstrating the ability to fly at Mach 3 – it was believed this was no longer sufficient to penetrate the Soviet defense system. Progress in Soviet anti-aircraft missiles would have nullified the bomber's ability to reach its targets, even at the high speeds and altitudes at which the B-70 was designed to operate.

But difficulties for the XB-70 kept mounting, its complex flight systems, the heat generated by high speeds, cost over runs, political resistance and a new growing reliance on ICBMs, intercontinental ballistic missiles, doomed the Valkyrie. By the end of 1967 the remaining XB-70 was put on display at the Air Force Museum, Wright-Patterson AFB, Ohio.

It was not, however, the end of the line for the bomber. Experience in subsequent conflicts – from the Middle East to Asia – was to prove the value of the

An in-house variant of the North American Model XB-70 titled "Phase 2.5". It shows a different configuration of canards and wing platform and has four engines. These and the following photos were actually saved in the late 1960s from going to the dumpsters when North American Rockwell was cleaning house. (Author)

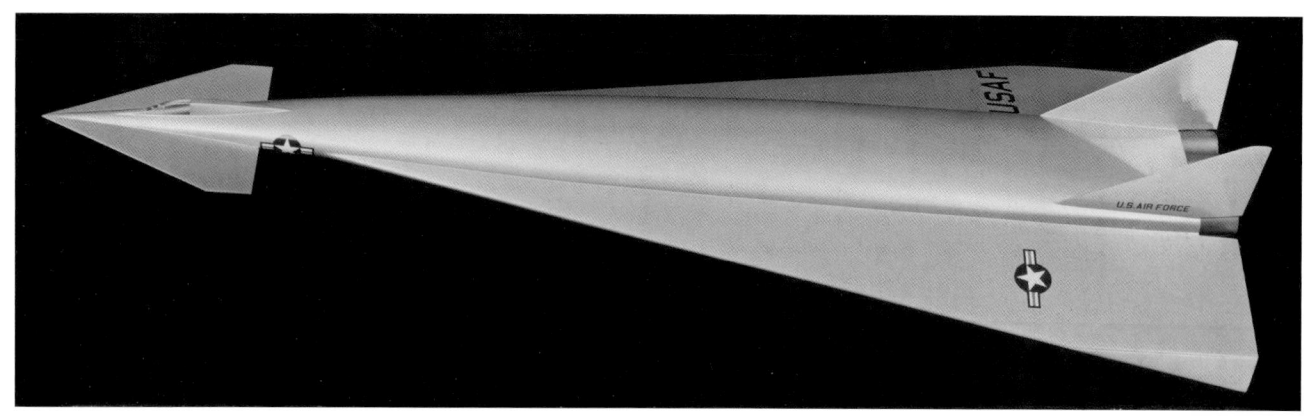

Above: *in-house wood model of the North American Model titled Phase 3 shows an elongated delta-wing platform with the wing tips in the folded position. The canards are similar to the previous model, but with a more pronounced clip. The verticals are canted outward and have four engine exhausts between them.* (Author)

Left: *In this stage of development, North American's in-house wood model shown has taken on recognizable characteristics of the XB-70.* (Author)

manned offensive aircraft. However, the latter's role would increasingly become that of delivering conventional explosives with pin-point accuracy rather than laying waste a city or region.

It is often thought that the Air Force bomber designation series ends with the B-70. Actually, one other number was allocated, the B-71, to a variant of the Mach 3+ Lockheed A-12 being considered for development as a strike bomber. Instead, a reconnaissance-strike version of the A-12 was ordered which used the sequence number -71. But instead of "B" for bomber, it became the RS-71. One urban myth attributes President Lyndon Johnson, in a speech about the new aircraft, accidently referring to it as the SR-71 and the prefix "SR" stuck for "strategic reconnaissance". The SR-71 Blackbird first flew on 22 December 1964.

Convair Aerospaceplane

During WWII the Germans conceived an idea of a long range bomber rocketed into sub-orbital space which would glide along the fringes of earth's atmosphere to its target. One such program was the Silbervogel (silver bird). This actually foreshadowed boosted lifting bodies, winged spacecraft and re-entry space vehicles. By the early 1950s Bell consulted with one of the former German engineers and developed the BoMi (bomber/missile), designated MX-2276. By the late 1950s Boeing started working the concept with their proposal, the X-20 Dyna-Soar.

In 1957, the Department of Defense issued a Study Requirement (SR89774) to initiate the research and planning of a reusable space booster. One of the results of this SR was the USAF Aerospaceplane, the first major attempt to come up with a large-scale logistical aircraft with the ability to fly into space and return to Earth using lifting body reentry techniques. This was – and remains – an eminently desirable goal.

By 1959, the idea of a reentry vehicle had evolved and was now termed the Recoverable Orbital Launch System (ROLS). ROLS was to be a large-winged single-stage vehicle capable of taking off horizontally from conventional runways, reaching low Earth orbit, and returning to base, landing in the same conventional manner.

The unusual thinking behind this early concept, however, had to do with its fuel system. It was a complex multi-phased propulsion fuel system devised to reduce the gross takeoff weight by not having to carry the oxidizer for its hybrid rocket system. It worked something like this: as the aircraft lifted off into higher altitudes, using up its conventional engines and fuel, it would extract the gaseous oxygen from the air around it. The oxygen would then be liquefied and mixed with the fuels to ignite and fly on its air-breathing hybrid rocket. Only one model has been found, so far, representing what is speculated to be Convair's Aerospaceplane design.

Speculation suggests this in-house model is the Convair Aerospaceplane. Although the identification is not certain, it appears close to one of the three-view illustrations of the Convair proposal. This angle shows the model's massive engines, providing sufficient thrust needed to get it into space. (John Aldaz collection)

An underside shot of the in-house model considered to be the Convair Aerospaceplane shows the curvature of its large delta wing. (John Aldaz collection)

An illustration of the North American M-3000 airliner concept, which used the XB-70 as a basis for its design. This concept appeared at a time when the American aerospace industry was considering Mach 3 as the next logical step in air travel. *(George Cox)*

Swords into Plowshares

Advanced military development can eventually trickle down to civilian industry. In the case of high-speed, long-range strategic bombers, this technology would eventually go to commercial airliners. Earlier examples of this process came from WWII. Britain's Lancaster bombers underwent the transition to civilian life as the Lancastrian. In America, the Boeing B-17 Flying Fortress transformed into the 307 Stratoliner and the Boeing B-29 Superfortress evolved into the Model 377 Stratocruiser.

For any aircraft manufacturer, it was a most logical step to repurpose knowledge gained in the research and development of military aircraft, applying it to the civilian market once the military deemed such applications as no longer classified. This, of course, opened other markets for aircraft manufacturers to ply their wares.

The airline business benefitted from the improved military technologies in airframe construction and power plants to give airlines safer, stronger, faster, and more efficiently operating aircraft. Throughout the 1950s and 1960s, the military had developed large supersonic aircraft with the likes of Convair's Mach 2 B-58 Hustler, North American's Mach 3 XB-70 Valkyrie, and Lockheed's Mach 3.2 A-12 Blackbird. As a result, the possibilities of creating a large fleet of commercial SSTs (supersonic transports) was seemingly within reach.

In 1962 when it was announced that the Anglo-French SST project, named Concorde, was to be given the green light, other nations began developing designs of their own. The USSR's Tupolev Design Bureau started work on its Tu-144 (allegedly drawing heavily on Concorde data), and in America, NASA began the SCAT (Supersonic Commercial Air Transport) program study.

When the American air carrier Pan Am placed its order with Concorde it had a similar 'wake up call' like the launch of Sputnik to the US airline industry. On 5 June 1963, President John F. Kennedy announced the authorization of the SST program, and the Federal Aviation Agency issued a Request For Proposal (RFP) for an SST design. The request was sent to three airframe manufacturers — Boeing, Lockheed, and North American, and to three engine manufacturers — Curtiss Wright, General Electric, and Pratt & Whitney.

While the Concorde was designed to fly 100 passengers at Mach 2 (1,250mph [2011.68km/h]) across the Atlantic, the Americans would aim for a much larger, 250-300 passenger airliner able to fly much longer routes. This greater load-carrying capability and longer range would make it a much more attractive commercial proposition to airlines. It would fly faster than Concorde, at about Mach 2.7. However, this would bring a penalty

The Bombers Chapter 2

in terms of complexity. Concorde's success hinged on it being at the limit of conventional technology in terms of proven engines and a largely aluminum airframe. The US initiative was to lead to some fascinating projects. Fortunately some artwork and a handful of models have survived to this day.

Below: Unknown Republic jet airliner concept model from about the late 1950s depicting a small aircraft being considered for production. *(Cradle of Aviation Museum)*

Above: Detail of an unknown Republic jet airliner concept model from the late 1950s similar to the one in photo at left, but exhibiting a different color scheme. *(Cradle of Aviation Museum)*

In-house wood model of an early North American concept for their first proposal of the SST based on the company's XB-70 Mach 3 bomber. *(George Cox)*

Republic Proposals

Although high Mach numbers were still out of reach for commercial airliners, the airline industry did benefit from other advancements in airframe and propulsion designs. Through most of the 1950s the jetliner had not yet arrived, and propeller airliners were more the norm. Using Republic as an example, here are some designs offered which included the promise of jet flight and the benefits of advanced prop aircraft. Republic's chief designer Alexander Kartveli proposed commercial jet airliner concepts as well as Republic's civilian Rainbow variant of the XF-12 design.

The Republic RC-2 Rainbow was a good example of converting a military airplane into an airliner at the beginning of the Cold War. This 1/72-scale wood model, built in Republic's model shop at Farmingdale, New York, sits on a lightning bolt stand used for factory P-47 models. This un-retouched photo was used by company artists to create paintings of the RC-2 as it might have appeared in flight. *(Republic photo via Mike Machat)*

A four-engine V-tail jet airliner concept was developed by Republic after the XR-12 program was cancelled. The close family resemblance can be seen in the nose of this in-house model. (Cradle of Aviation Museum)

The Bombers   Chapter 2

A large plastic model made by Boeing shows its 2707-100 with the variable-geometry wings in the fully swept-back configuration for supersonic flight. Eventually this version was lengthened and canards added for better stability to become the 2707-200. *(Ron Monroe collection)*

Boeing 2707

Boeing began studies of SST concepts as early as 1952. By the 1960s it offered a variable-geometry wing version, the 2707-100 and a fixed-wing version, the 2707-300, eventually settling on the latter because of the complexity and added weight of the swing-wing design. With this, Boeing duly won the competition to be the producer of the first U.S. SST airliner. An impressive mock-up was constructed but by 1971 (two years after Concorde's first flight) the world had changed. Fuel costs had escalated dramatically as a result of Middle East conflict and would never return to their earlier levels. At the same time, environmental issues such as noise, pollution and supersonic bangs (the shock waves that hit the ground from supersonic flight) had become much more significant public concerns. As a result the whole program was terminated.

Inboard profile of the Boeing 2707-200 SST shows the seating layout. This was a larger version of the 2707-100 but problems from the excessive weight of the wing-swinging mechanism could not be solved, so Boeing went to the next version with fixed wings, the 2707-300. *(Scott Lowther)*

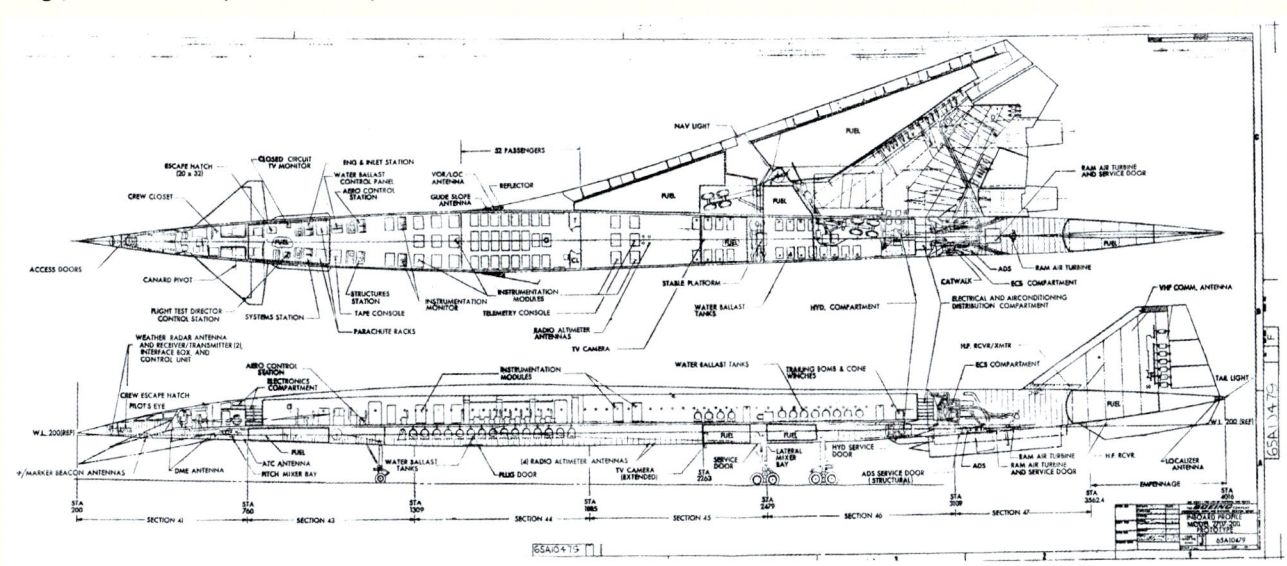

83

Once Boeing was awarded the government contract to build the SST, it settled on this 2707-300 model as its final design. This model has fixed wings and was the longest and heaviest of all the versions. *(San Diego Air and Space Museum)*

Lockheed SST

What better manufacturer to build America's first SST than the company that created the world's first Mach 2 jet fighter, the stunning F-104 Starfighter and the Mach 3+ SR-71 Blackbird? Lockheed's Advanced Design Group, headed by the legendary Kelly Johnson, seemed like a shoe-in to develop a supersonic airliner, but as noted it was Boeing – a builder of large bombers and transports – that won the competition. However, Lockheed's Model 2000 looked every bit as capable of delivering the goods.

An in-house model of the Lockheed L-2000-1 concept for their SST. This company was chosen along with Boeing to continue development but eventually was eliminated from the competition. Their new design could seat 170 to 200 passengers depending on the layout. *(San Diego Air and Space Museum)*

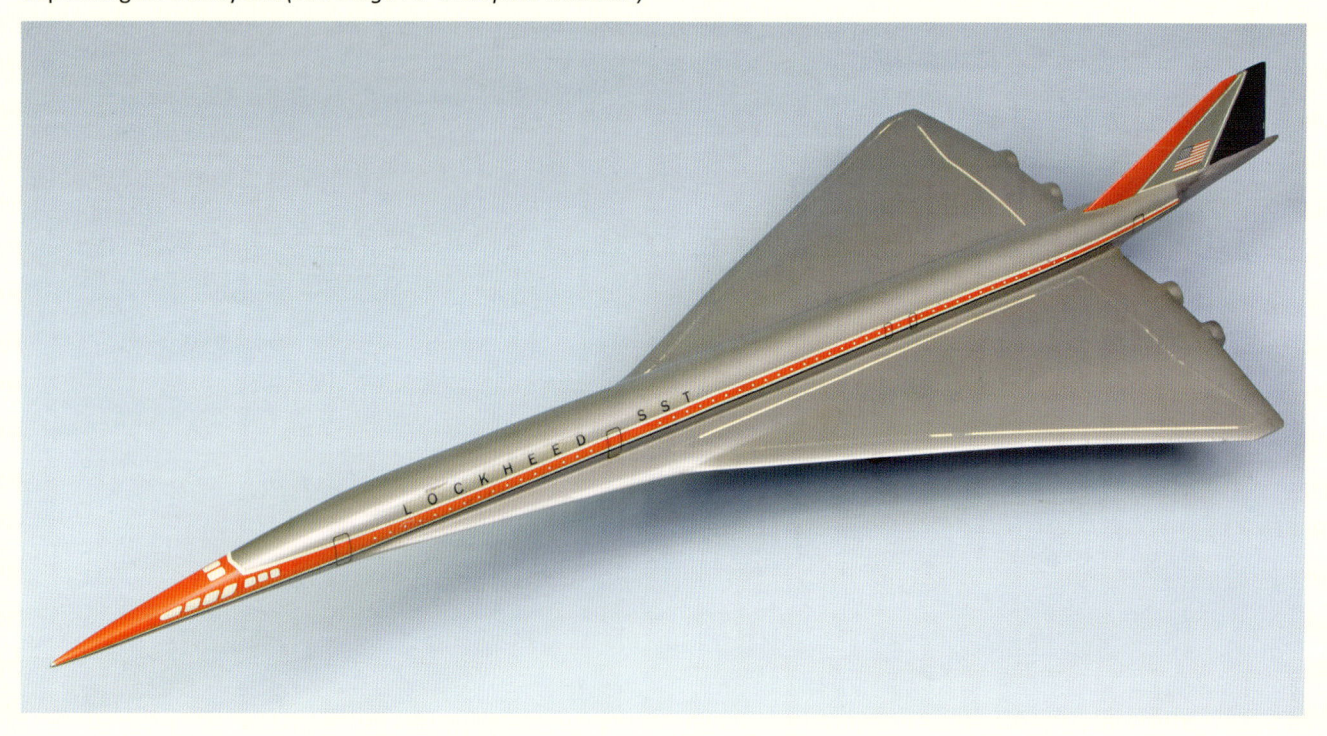

The Bombers Chapter 2

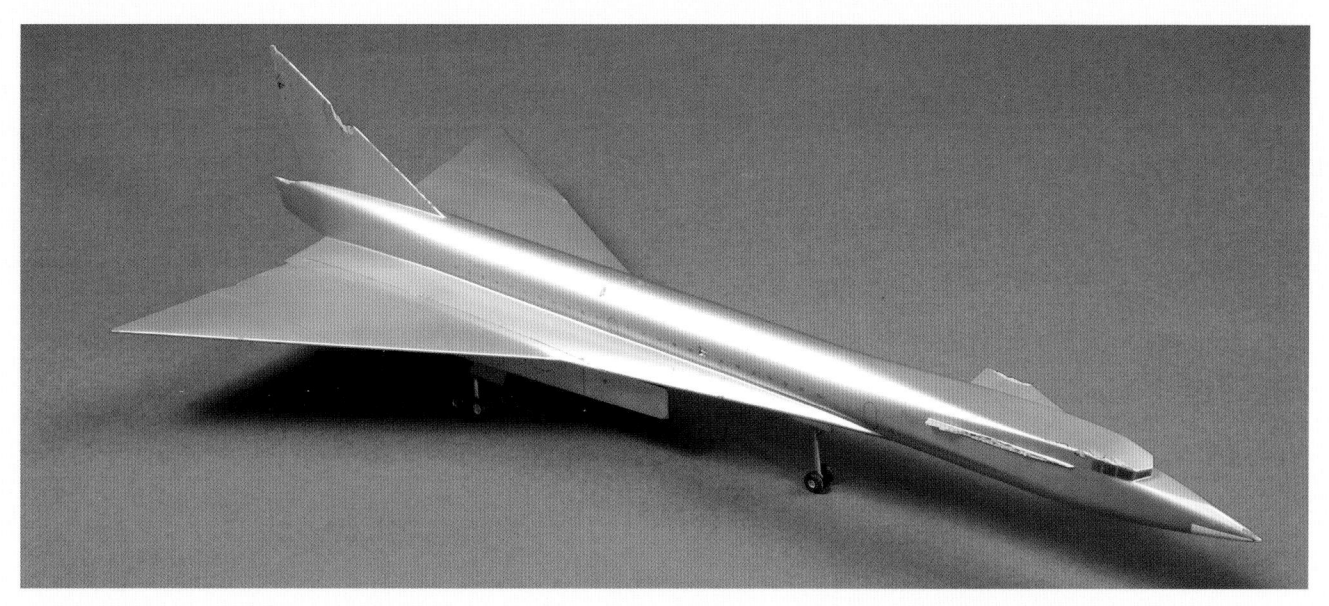

Although Douglas was not even in the competition, it conducted its own studies of the SST. This particular in-house wood model was destroyed during the 1994 earthquake. It is one of two known variants used for the study of different nose configurations and intake designs. *(Author)*

Douglas Model 2229

Douglas also explored designs beyond the initial SST specification, looking at Mach 3 possibilities (as did North American Aviation). Douglas' proposal had a fuselage length of 200ft (61.0m), a wingspan of 100ft (30.5m) and when the wingtips were fully extended a gross weight of 400,000lb (181,437kg). Like the XB-70, the outermost 20ft (6.1m) of each wing could be folded down for supersonic flight. This feature increased the favorable pressure distribution beneath the wing created by the leading-edge shock wave and supersonic compression alongside the tapered propulsion box of the lower fuselage.

It has happened numerous times throughout commercial aviation history, but the so-called "need for speed" has often been trumped by the sheer reality of contemporary economics. Even an airliner such as Convair's elegant 990 Coronado which flew at the edge of the speed of sound was outstripped economically by the Boeing 707 and Douglas DC-8. Despite being a remarkable technological success – able to cruise for three hours at speeds beyond the 'dash' capability of most contemporary fighters – Concorde was a commercial failure. After spending nearly three decades in airline service it never earned a profit. With no other SST's on the drawing board to this day, Concorde was both the first and (for the foreseeable future) the last supersonic airliner to carry fare-paying passengers. Only plans, artwork and models remain as testament to the quest for a successor.

Close-up detail showing the cockpit and nose for another of the Douglas SST studies. Again, the damage shown here is from the 1994 earthquake in California. *(Author)*

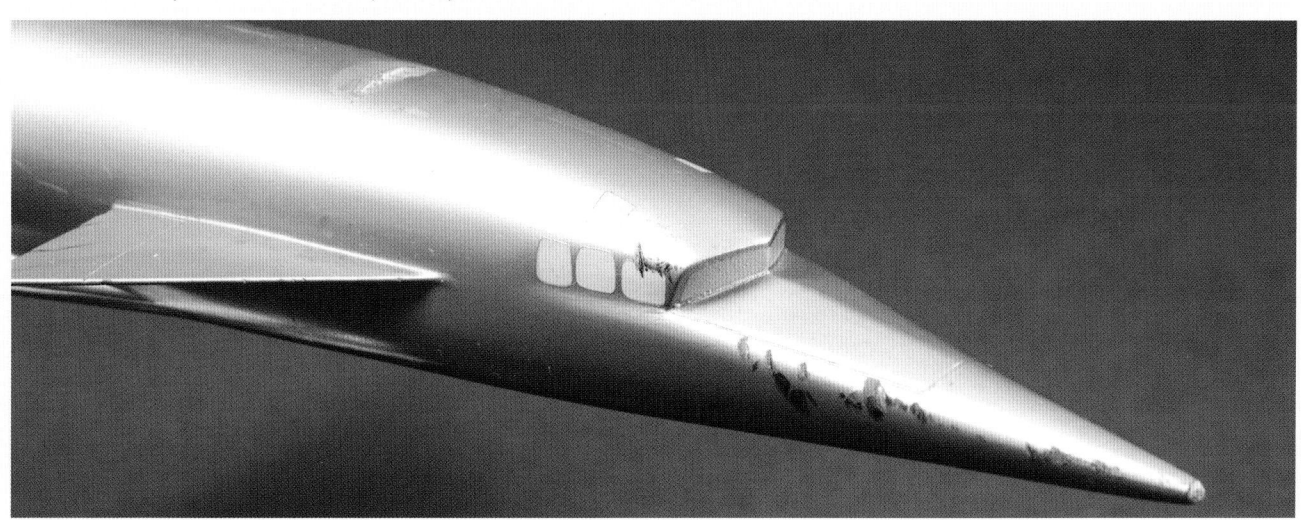

Chapter 3
The First Jets

Artist Mike Machat depicts the first military flight of the Bell XP-59 over Rogers Dry Lake at Muroc, California (now Edwards Air Force Base), late in the afternoon of 2 October 1942. An aviation artist goes to great lengths in researching his subject matter in order to depict the correct lighting, weather conditions, and even the flight attitude of the final approach. (Mike Machat)

Renowned aviation writer Robert Serling is quoted as saying, "Progress in aviation took place one nut and bolt at a time." While true in the sense that improvements in aeronautical design and aircraft performance are incremental, there are moments in aviation history when a significant paradigm shift takes place. Such a time was the advent of the turbojet engine.

Development during World War II

Again, it is useful to take a quick look at the history of jet development in order to understand our subject better. The book covers the period 1945 to 1965, by which time the jet age was well underway. However, the first jets emerged in the late 1930s and developed rapidly during World War II. The Germans worked on several designs for jet and rocket-powered fighters and even for a number of early jet bombers. They had flown the first-ever jet aircraft, the Heinkel He 178, in August 1939. They went on to put a handful of designs into production, the most successful of which was the twin-jet Messerschmitt Me 262 fighter, even though this had to be rushed into service in the desperate closing stages of the war.

# The First Jets	Chapter 3

Special Hobby produced a 1/48-scale plastic model of the Gloster E.28/39, Britain's first jet powered aircraft. The designation of E.28/39 comes from a government specification known as E.28/39 and it stand for the 28th "Experimental" specification issued by the Air Ministry in 1939. This aircraft is also known as the Gloster Pioneer, Gloster Whittle, and Gloster G.40. (Allen Hess model)

The Allies were a step or two behind the Germans in developing operational jet-powered aircraft, even though the British had arguably taken the lead in engine development. The British engineer (and incidentally, outstanding Royal Air Force pilot) Frank J. Whittle had earlier produced ground-breaking papers (studied by his German counterparts) on the subject of the gas-turbine engine as a means of aircraft power. It took him time – and a great deal of perseverance – to raise official interest in his ideas, but by the early 1940s, Whittle had developed an engine suitable for flight. His 623lb (282.6kg) centrifugal-flow Power Jets W.1 could deliver 855lb (3.8kN) of thrust and was installed in a specially constructed Gloster E.28/39 trials aircraft. It flew for the first time on 15 May 1941.

As the war progressed the Germans, driven by desperate need, produced many more engines than the British, and introduced jet-powered fighters into squadron service well ahead of the allies (for whom only the RAF's Meteor would see operational service, intercepting V-1 flying bombs). However, the British engines proved far more reliable and could run for far longer between overhauls. This was down to two things. One was the fact that the British employed nickel alloys for the hottest parts of the engine, whereas the Germans had to use stainless steel. The other was the fact that the British chose the much more robust centrifugal compressor rather than the German's axial flow arrangement. Unlike the latter, centrifugal engines do not suffer from compressor stalls. When this happens the airflow through the engine can get reversed and result in a flame-out.

Indeed the early British engines proved remarkably reliable, even compared with the piston engines they replaced. The need to attempt 're-lights' in flight was virtually unknown. As a consequence, after the war, the tiny Vampire went on to become the first jet aircraft to fly the Atlantic – on its single engine!

During 1941, General Henry H. "Hap" Arnold, commanding General of the USAAF, happened to be in England and witnessed the first test flights of the Gloster Whittle. Upon his return to the States, he arranged immediate action to procure an agreement with the British and have this new engine built in the United States by General Electric. General Electric no sooner had the engine built and running then it started making its own improvements and eventually came up with a variant called the 1-A. It was this engine that powered the first American jet aircraft, the Bell XP-59A.

Hobbycraft produced a plastic and resin 1/48-scale model of the Bell XP-59A. The development of this aircraft was so secret that in order to throw off suspicion that it was a jet, a wooden propeller was attached to the nose while being transported to Rogers Dry Lake for testing. (Allen Hess model)

The development of the first American jet was ranked top secret, just below the most secret of all projects, the development of the atomic bomb. Bell Aircraft was chosen as the airframe developer partly because of its proximity to General Electric in New York.

The Bell XP-59A was constructed specifically to take on a pair of 1-A engines. Its first flight took place on 1 October 1942. The initial results were disappointing with poor flight-handling characteristics and very unreliable power plants: hardly surprising as this was unproven technology in a hastily constructed airframe.

Although the outcome of those initial flight tests were below expectation, the USAAF placed an order for no fewer than 80 P-59s – even before the YP-59 preproduction models had flown. This seems to have been a compulsive response to the wartime pressure to field an operational jet-powered aircraft. The USAAF had all the intentions of making the P-59 a viable warplane, but with the latest piston engine fighters reaching the peak of their development, it was not up to the job. Instead, the P-59B became a trainer for the future jet pilots of America.

Even before the Whittle jet engine had arrived in the states, U.S. companies such as Pratt & Whitney, Northrop, and Lockheed were contemplating the use of turbines, turboprops, and axial-flow turbojet engines. In March 1942, Lockheed Aircraft was the first to propose its own engine and aircraft combination. It had designed a 5,100lb (22.67kN) axial-flow turbojet engine, which was named the L-1000, and was planning to put two of them in its new fighter jet aircraft, the L-133.

The proposed L-133-02-01 was a single-place jet fighter weighing in at 18,000lb (8164.7kg) gross weight. It was a highly futuristic aircraft with a predicted speed of 612mph (984.9km/h) in level flight, and 710mph (1142.6km/h) in a dive. But maybe this was just all too new for the USAAF, particularly when America was stepping up production of tried and trusted piston-powered designs. The L-133 with its two L-1000 engines was turned down, but the idea of designing an airframe to match a specific power plant was forward thinking that would be revisited in the post-war era.

This Lockheed model of the L-133 model is made of resin. The two L-1000 axial-flow engines are located just behind the pilot's seat with the air intakes at the nose. This made for a very long air intake run. By today's standards the intake looks too small. (Allen Hess model)

This angle shows the streamlined form of the Lockheed L-133. An impressive proposal for its time, this resin model shows the smooth contours of this tailless design with blended wings and canards at the forward fuselage. (Allen Hess model)

As the war progressed aircraft manufacturers started to come up with their own jet designs and offer them to the military. These first jets did not spring from military competitions. That wouldn't start until 1945 when the Air Force would call for the new breed of interceptors. Aircraft makers independently worked up designs based around the new jet power-plant knowing that this would be the future of aircraft production. One of the first successes was the Lockheed P-80 Shooting Star, which flew in October 1943. It entered service too late for World War II but did see action in the Korean conflict.

Other fighters soon followed. Republic produced the F-84 which flew in 1946. And North American Aviation followed up in 1947 with the first American swept-wing jet, the F-86 Sabre. These early jets took the USAAF firmly into the jet age and placed the United States at the forefront of fighter capability.

However, while companies were working on the first generation of jet fighters, the limitations of the early axial flow turbojet were giving them headaches. These early engines were low on thrust and reliability, high on fuel consumption and maintenance. As a consequence experimentation of mixed power-plants were tried out.

The Ryan XFR-1 Fireball, a composite of piston engine and turbojet, flew in 1944. The Convair XP-81(see sidebar "Composite Fighters") a composite of turboprop and turbojet engines came next, flew in 1945 but never progressed beyond the prototype stage.

The Ryan XF2R-1 Dark Shark model represents an aircraft to be one of the best performing composites. This highly Modified FR-1 never got the chance to go into production, for it was dumped in favor of pure jet aircraft. *(San Diego Air and Space Museum)*

Composite Fighters

Some of these composite proposals planned to convert existing airframes while the others were completely new aircraft. The idea was to take advantage of both worlds, the fuel efficiency of the propeller power plant in combination with the jet's ability to function at high speed and to be used for additional power boost. Thus, a fighter's performance could be enhanced at takeoff by using the jet engine to climb quickly to altitude, then using its additional thrust for a burst of speed to engage the enemy. However, these were mainly short-lived experiments. These interim designs were eventually overtaken by aircraft powered by new and more efficient and more powerful turbojet engines.

Nonetheless, the US Navy was especially perseverant with the concept. The Navy liked the idea of high-performance sea-going aircraft. The drawback was that the early jets required long takeoff and landing runs, consumed enormous quantities of fuel, and, most significantly, had very slow response times if and when the pilot opened the throttle: all fairly undesirable characteristics when operating off an aircraft carrier! Hence, the Navy took a step back and looked at composite power plants as a solution. The Ryan FR-1 Fireball and its successor, the XF2R-1 Dark Shark, were the first results. (Those aircraft are covered in more detail in the Chapter Four).

Convair-Vultee XP-81 Silver Bullet

In the meantime the USAAF ordered two prototypes designated XP-81 on 11 February 1944. This was to be a single seat, long-range escort fighter that combined a single General Electric TG-100 turboprop engine in the nose with a GE J33 turbojet in the rear of the fuselage. Since the turboprop was not ready for flight tests, a Rolls-Royce Merlin engine from a P-51 was installed. The flight tests, held at Muroc's Rogers Dry Lake, were concurrent with the Navy's testing of the Dark Shark. The Army Air Force liked the results from the Dark Shark better than its own XP-81, and consequently the USAAF took up the Dark Shark once the Navy expressed no further interest in it. The XP-81 was cancelled.

After the TG-100 turboprop was finally installed in the XP-81 and flight testing resumed, Convair discovered that the turboprop engine would not produce its designed power, giving the same results as the earlier Merlin-powered configuration. Testing ended in 1947. *(John Aldaz collection)*

The rear view of the Convair XP-81 wood model shows the exhaust port of the GE J33 turbojet, which was placed in the rear of the fuselage. *(John Aldaz collection)*

North American FSW P-51

Not all composites were entirely new aircraft. Experiments were done using existing inventory. Douglas Aircraft considered augmenting the performance of its A-26 Invader by adding a jet engine to the twin-engine piston-powered bomber. The experimental XA-26F was powered by piston engines with four-blade propellers, plus a turbojet placed in the rear of the fuselage.

Perhaps the most adventurous idea came from North American Aviation. This company sought to exploit the highly successful P-51 Mustang airframe by utilizing two new developments, the jet engine (putting a Westinghouse jet in the rear fuselage to boost performance in climb and combat) and swept wings. Information on the latter had just been captured from the Germans. Although the proposed forward-swept configuration looks odd today (following sixty years of seeing wings swept backwards), back in 1945-1946, it wasn't clear which direction of sweep was best. Both offered the advantage of countering the effect of shockwaves at transonic speeds, and forward sweep actually offered some stability advantages. However, wind tunnel tests showed that aerodynamic twisting of the forward swept wing would lead to severe structural problems and so the plane was never built.

Top left: A closer look at the FSW P-51's mid-fuselage air intake for the jet engine that was mounted in the rear fuselage. *(George Cox)*

Top right: The underside of the Forward Swept Wing P-51 shows the tail pipe for the jet engine mounted in the lower rear fuselage. *(George Cox)*

bottom: North American considered putting forward-swept wings on its P-51 and adding a turbojet. This Al Parker model shows what the configuration would look like. *(George Cox)*

Artist's rendering of the Curtiss-Wright P-304-04 taken from the engineering plans shows the first of two tailless versions of their jet fighter. As of this writing, the images in this book are the only known existing P-304 illustrations. The P-304-04 utilized the open nose air intake for the General Electric TG-180 turbojet. (National Archives)

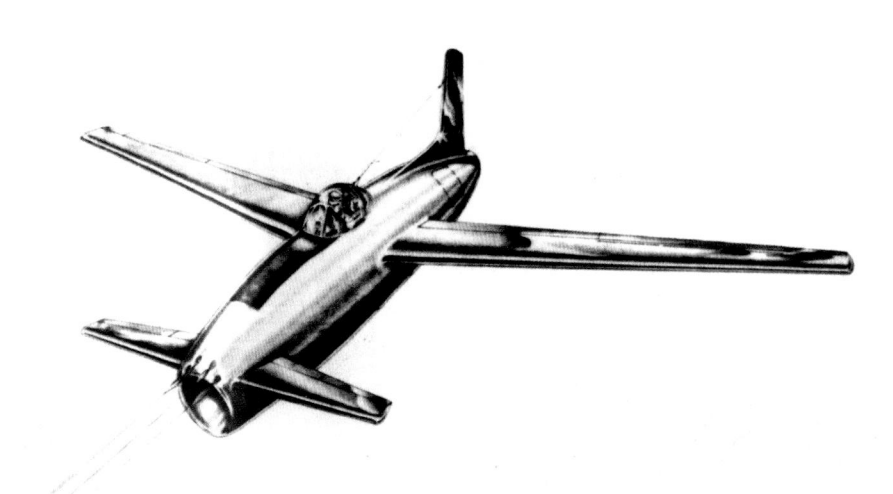

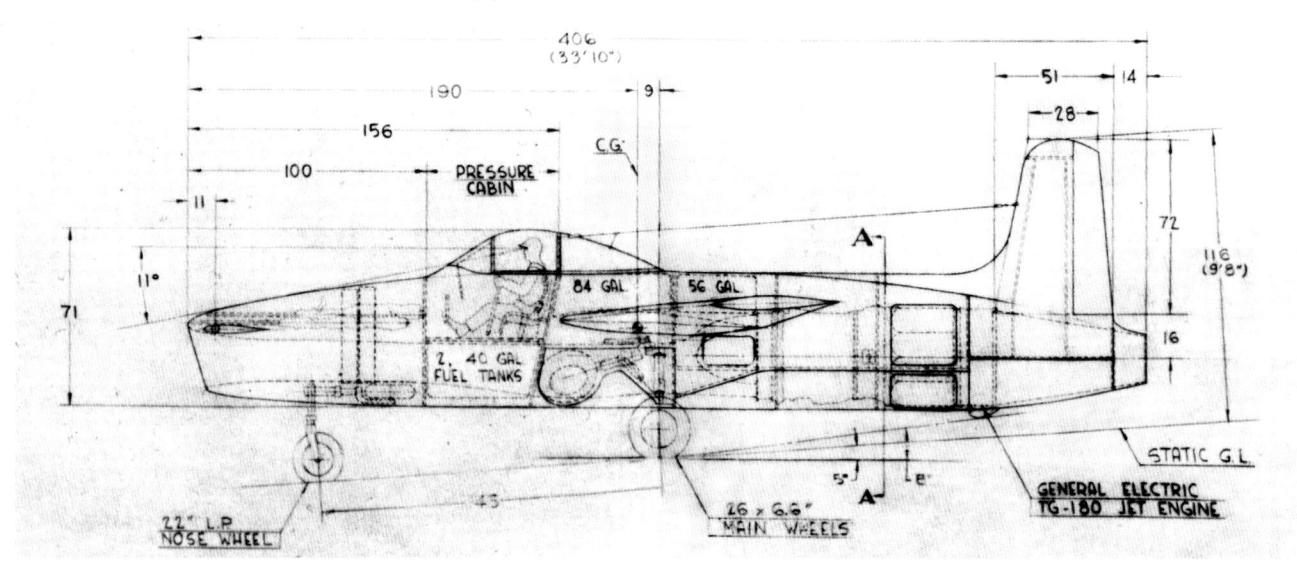

A side layout of the Curtiss-Wright P-304-04 shows the pilot, landing gear and engine locations. At just under 34 feet (10.4m) in length it was slightly longer than its predecessor, the propeller-driven XP-55 Ascender. The P-304-04 had a wingspan of 42 feet (12.8m), two feet longer than the Ascender. (National Archives)

The Curtiss-Wright P-304-04 and P-304-05

During WW II a number of propeller-driven aircraft were in development, competing for military contracts. As the war started to wind down, it became evident that turbojets would soon become the desired power source of the future. Consequently, some of the airframes like that of the Douglas XB-42 were easily converted from props to jet propulsion. Others needed a little more reconstruction. Such was the case with the Curtiss-Wright XP-55 Ascender.

On 27 November 1939, the United States Army Air Corp issued its "Request for Data R-40C" calling for unconventional fighter aircraft designs. The idea was to encourage aircraft manufacturers to depart from conventional thinking in order to come up with something different in aircraft design. Three aircraft manufacturers won the right to compete with what were then considered fairly radical proposals: Vultee with the XP-54 Swoose Goose, Curtiss-Wright with the XP-55 Ascender and Northrop with the XP-56 Black Bullet.

The Curtiss-Wright XP-55 Ascender (Curtiss-Wright CW-24) was a highly unusual design for its time. The aircraft had swept wings with two vertical tails, a rear-mounted engine, tricycle landing gear and a canard configuration. This pusher type was designed for the Pratt & Whitney X-1800 engine, but Pratt & Whitney could not complete the new engine and the program was canceled. The XP-55 was then redesigned to take the inline, liquid-cooled Allison V-1710 (F16) engine rated at 1,000hp (746kW). The XP-55 first flew on 13 July 1943. The performance of the XP-55 was plagued with a number of problems, and by 1944 the project was canceled.

Curtiss-Wright was, of course, disappointed that its Ascender did not do well. But Curtiss felt the research and data gathered from the XP-55 trials indicated that it was worth a second try. Now that the "jet age" had dawned, it seemed very logical to resurrect the Ascender as a jet-plane proposal.

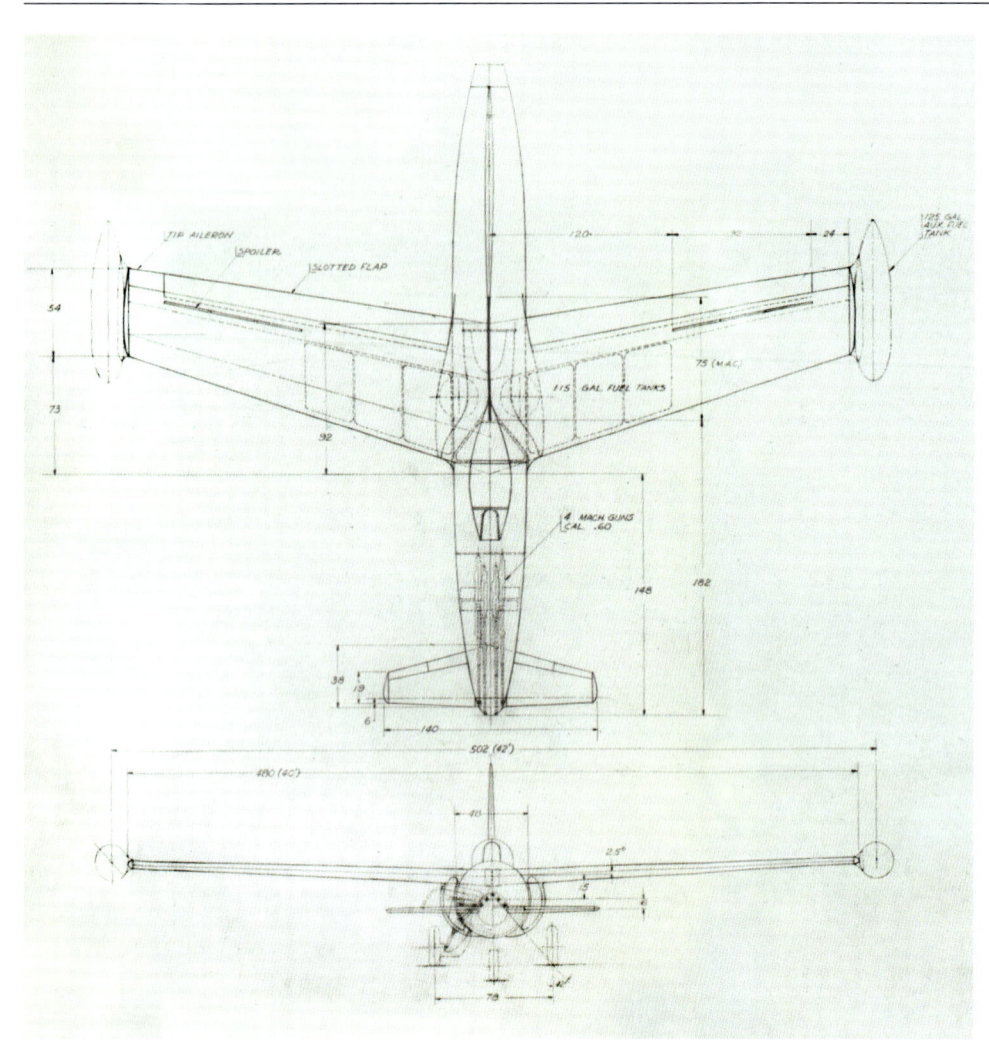

Front and top view of the Curtiss-Wright P-304-05 shows the added fuel cells inside the wing, the four 0.60-caliber machine-gun placements, and the 15-degree sweep of the wings. (National Archives)

Wings72 produced a vacuform model of the Curtiss-Wright XP-55 "Ascender." For its time, the aircraft was considered quite radical, with its swept wings, vertical tails on the wing tips, rear mounted engine and a canard configuration. A special feature of the XP-55 was the ability to jettison the propeller in an emergency to prevent the pilot from hitting it during bailout. (Barry Webb model)

The First Jets   Chapter 3

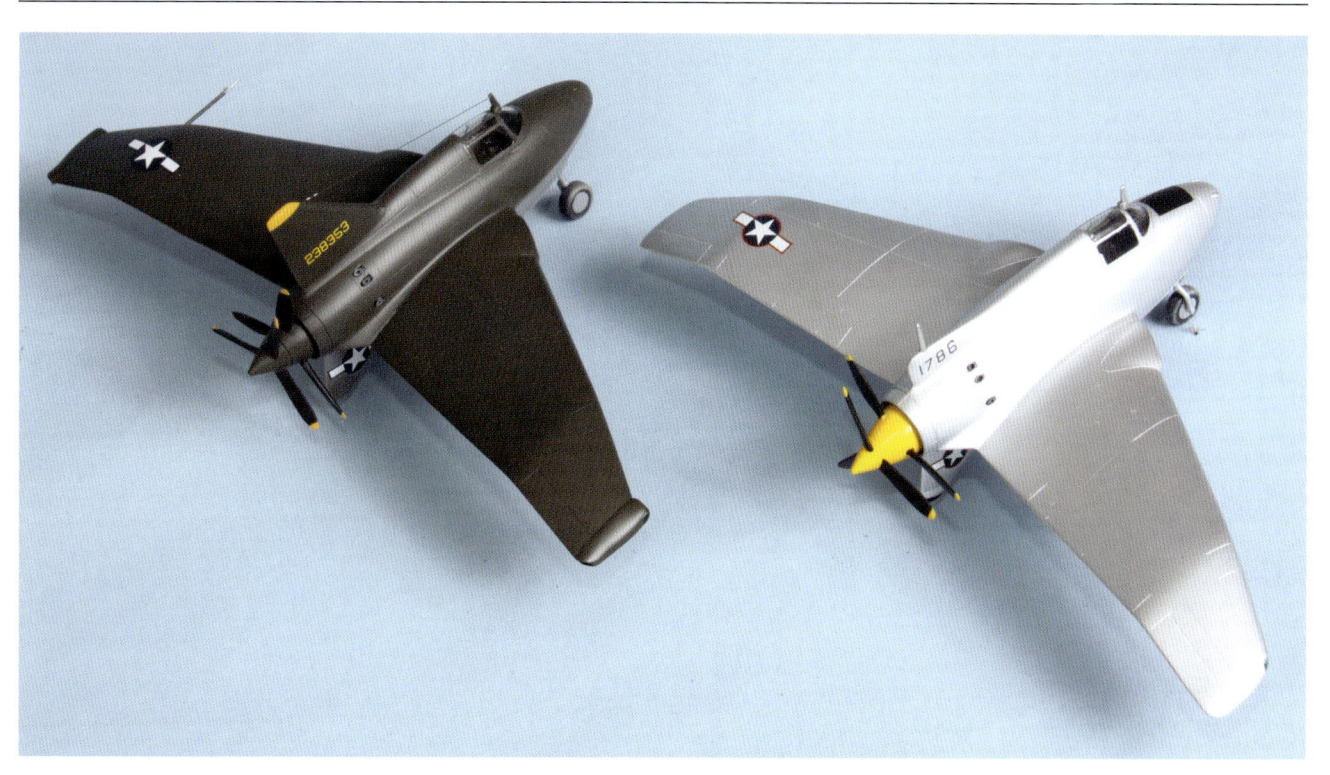

Two prototypes of the Northrop XP-56 "Black Bullet". The first silver version experienced pronounced yaw during flight so a larger vertical was installed. But it was destroyed during high-speed taxi tests when a tire blew out. The model on the left has the larger fin and wing tip bellows for rudder control. Both were designed around the Pratt & Whitney R-2800-29 radial, 2,000 hp (1,492 kW) engine driving contra-rotating propellers. MPM 1/72-scale model. (Barry Webb model)

The Vultee XP-54 "Swoose Goose" actually beat the Curtiss XP-55 Ascender and the Northrop XP-56 Black Bullet in competition. The first of two prototypes, represented by this resin model, flew in 1943. But eventually their performance did not fulfill the required specifications and the program was subsequently canceled. (Allen Hess model)

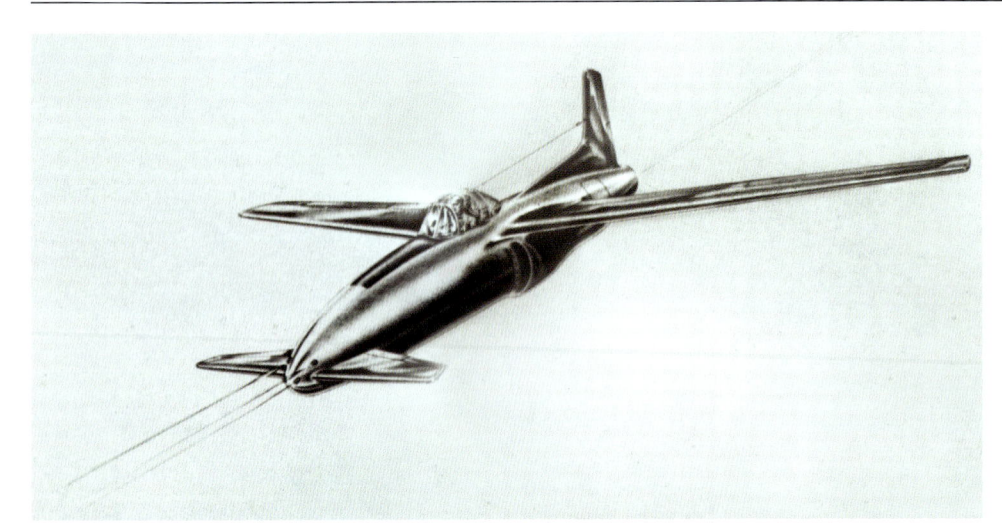

Artist's rendering of the Curtiss-Wright P-304-05 taken from the engineering plans shows differences from the P-304-04. The P-304-05 has a closed nose with the air intakes now under the leading edge of the wings and the wings are located higher on the fuselage. (National Archives)

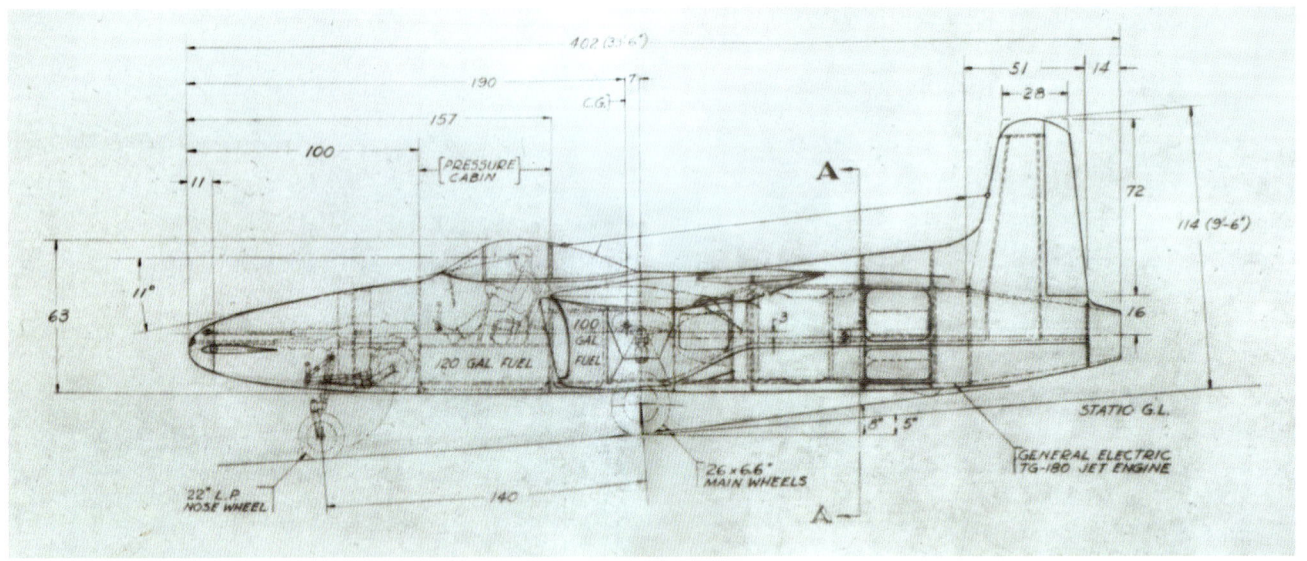

When comparing the P-304-05 side view to the XP-55 Ascender, the dimensions are approximately the same but the P-304's bubble canopy gave much improved pilot visibility. Curtiss-Wright was hoping that the military would accept either of the two P-304s on the basis that a lot of preliminary work had already been done with the XP-55. (National Archives)

In a report dated 31 March 1945, the Curtiss-Wright Corporation, Aircraft Division, came out with its Model P-304-04 and Model P-304-05 proposals. The pitch utilized data and experience obtained from the XP-55 program and converted it into a jet. The introduction in this report sums up the problems that any aircraft manufacturer must face when considering jet aircraft production:

"Proposal P-304 described herein presents an engineering study of a single-engine, jet-propelled, medium-range fighter with special emphasis given to high-speed performance, maximum combat utility and safe operation. Although jet-propelled airplanes were rapidly coming to the fore as combat types because of their high-speed characteristics, several limitations are presently imposed on their potential maximum utility. The two outstanding limitations are fuel capacity and compressibility effects at high speeds. The former is due to inherent characteristics of the power plants, which require two to three times as much fuel for the distance traveled as do conventional engines. The latter effect imposes a restriction on the conventional airplane design around a jet engine, limiting it because of safety and increased drag to a speed lower than that which it would be capable of attaining if its design configuration could be altered.

"The P304 design is based on an unconventional configuration directed particularly toward surmounting the present obstacles in the way of safe high-speed operation. This design is based on the accumulated knowledge derived from the years of development on the type, the program which culminated in the XP-55 airplane. Development of the XP-55 has now reached a stage where its limitations and possibilities for future action can be clearly defined:

- The means of propulsion must be changed from propeller to jet in order to increase its speed performance by 50%, in accordance with present day demand.
- Stalling characteristics must be improved to the extent that the type is equal or superior to conventional types.
- Performance in other categories such as climb, rate of roll, takeoff distance, and landing speed must be improved considerably.
- Longitudinal control forces must be made desirable to the pilot through the entire flight range.

"It is firmly believed that all the objectives outlined above can now be attained within a reasonably short period, provided that suitable research facilities are immediately available when required."

In conclusion, the report states, "... this design represents a very attractive possibility of creating an airplane in the fighter category which has a better than average chance of operating at the high Mach numbers without disastrous effects. It also represents a logical continuation of the XP-55 program, on which much time and effort was expended, with a very good chance of 'cashing in' on the results."

Here are the specifications on the P304-04 and the P304-05:

Engine model	General Electric TG-180
Maximum rated thrust	4,000lb (1814.4kg) at sea level
Wing area	240sq.ft (73.1m^2)
Wingspan	40ft (12.2m)
Sweepback	15°

	P-304-04	P-304-05
Length	33ft 10in (10.3m)	33ft 6in (10.2m)
Height	9ft 10in (2.9m)	9ft 6in (2.89m)
Design gross weight with 450gal (1703.4l) internal fuel	12,800lb (5,806kg)	12,500lb (5670kg)
Wing loading	53.3lb/ft^2 (24.1 kg/m^2)	52.1lb/ft^2 (23.6 kg/m^2)
Gross weight with reserve fuel only	10,300lb (4672kg)	10,000lb (4536kg)
Wing loading	43.0lb/ft^2 (19.5kg/m^2)	41.7lb/ft^2 (18.9kg/m^2)
Gross weight including external fuel 700gal (2649.8l)	14,470lb (6564kg)	14,170lb (6427kg)
Wing loading	57.6llb/ft^2 (25.6kg/m^2)	56.5lb/ft^2 (26.1kg/m^2)
Combat range radius with 690gal (2611.9l)	500 miles (804km)	500 miles (804km)

Curtiss-Wright's proposal was trying to make the point that its new designs were a continuation and an incorporation of the data obtained from the XP-55 program, and that the monies spent on the XP-55 would not go to waste if only the military would just green light this program. In the end, this proposal was declined.

The Call for New Types of Aircraft

Toward the end of the war, the U.S. military identified three types of new aircraft needed to replace current inventory. All three required jet or rocket power. One was a rocket interceptor, able to get to altitude as quickly as possible and intercept incoming high-speed, high-altitude enemy bombers. The second was a penetration fighter, able to escort bombers over enemy targets, and the third was an all-weather fighter capable of seeking out targets on its own, day or night, on the ground or in the air.

By the end of the war, U.S. aircraft manufacturers were already coming up with designs and proposals in an effort to meet the requirements of these three new specific types of aircraft. This was a time of exploration and development; for airframes and for power-plants, the technological envelope was being pushed. Inventing airframes that could withstand the rigors of jet flight was not as difficult as inventing the power-plant that could propel them. It was not uncommon for an aircraft to be proposed and developed but ultimately prove incapable of meeting its expected performance because the appropriate power-plant was either not ready or not up to the mark.

All this development and research was done as a business. Companies had to survive the ups and downs of the economy and the politics. Projects would be started, changed, canceled, then started again. Even strong aircraft companies could suffer the economic consequences of this roller-coaster ride.

As we look at some of the different companies through proposal models and artwork, keep in mind that the picture is incomplete. Companies kept project records for each program. Sometimes, however, after the designs and ideas were nullified by more advanced technology or by new military requirements, mounds of paperwork that seemed to have no more significance were thrown away. It was an "out with the old and make way for the new" mentality.

Models were tucked away on a shelf or stored in their shipping boxes, locked away in a closet. Some models only exist today because someone had "decorated" an office or the company library area with them. Some eventually made their way to museums, while others were taken home as keepsakes by employees or even given to kids as playthings. Decades later these still turn up occasionally in an antique store, at a swap meet or internet auctions.

Curtiss-Wright: First to Fall

Curtiss-Wright, an amalgam of the first two famous names in aviation history, had been very successful during World War II with its P-40 Warhawk series and the SB2C Helldiver. But it had also gone through a series of recent failures for the Army Air Force with such projects as the XP-46, P-60, and XP-62, and for the Navy with the XF14C, XBTC-2, and XBT2C-1. With the massive cancellations and layoffs that had descended on so many of the major U.S. aircraft manufacturers after the war ended, Curtiss-Wright was on the verge of closing down and would soon lose its footing in the aircraft manufacturing business.

On 23 March 1945, the USAAF announced a competition for an all-weather fighter bomber, looking for the successor to the Northrop P-61 Black Widow night fighter. Bell, Consolidated-Vultee, Northrop, Douglas, Goodyear, and Curtiss-Wright all submitted proposals. Curtiss-Wright had earlier submitted a design for the attack role of a similar configuration to the Model 29, called the Model 100. This was subsequently designated the XA-43, but when that project was canceled, Curtiss kept the same airframe arrangement and adapted it to the all-weather night fighter requirements. The new design was re-designated XP-87 and given the name Blackhawk.

The XP-87 was powered by four 3,000lb (13.3kN) thrust Westinghouse XJ34-WE-7 turbojets mounted under the wings paired in two large pods – a unique layout for a fighter. It was a large, all-metal, cantilever, mid-wing aircraft, with the pilot and radar operator seated side-by-side in a large cockpit. The gross weight of 49,000lb (22,226kg) made it one of the heaviest Curtiss aircraft ever built. Originally it was to be armed with a pair of .50in (12.7mm) machine guns mounted on automatically operated nose and tail turrets, along with internally mounted rockets. Later, the guns were changed to four 20mm cannon housed in a nose turret on a moveable platform, which allowed varied firing angles ranging from zero to 90°. This turret was to be developed by the Glenn L. Martin Company of Baltimore.

The prototype XP-87 was finally ready for its first flight at Rogers Dry Lake (Muroc) in March 1948. It took to the

The proposal brochure for the Curtiss-Wright XP-87 Blackhawk includes this cutaway view illustrating the placement of the fuel tanks, the twin-engine G.E. TG-180 (J35) and landing gear. (National Archives)

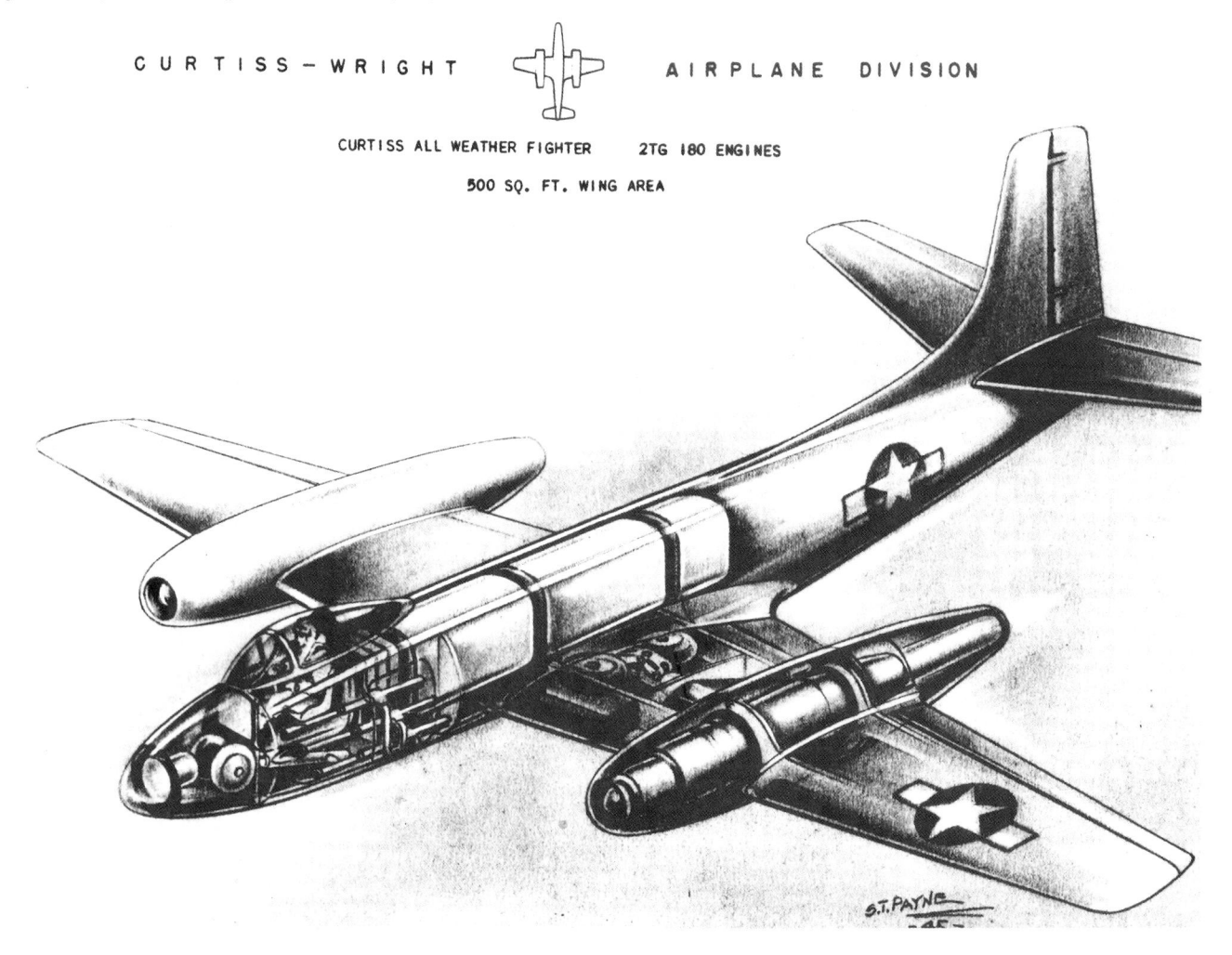

The First Jets   Chapter 3

An illustration from the Curtiss-Wright XP-87 Blackhawk proposal brochure depicts the ease with which the guns and the G.E. TG-180 (J35) engine can be accessed and serviced. (National Archives)

This in-house metal model prototype of the Curtiss-Wright XP-87 Blackhawk all-weather night-fighter was in competition to replace the Northrop P-61 Black Widow. Notice the difference in design change between the twin-engine concept drawings and this four-engine model. (John Aldaz Collection)

air for the first time on 5 March, proving to be underpowered and experiencing buffeting problems at the higher speeds. During the testing of the XP-87's development, the military changed the "P" designation on pursuit aircraft to "F" for fighters. And even with the teething problems of the XP-87, on 10 June 1948 the USAF placed orders for fifty seven F-87A fighters and thirty RF-87A photo-reconnaissance aircraft.

The new contract would breathe new life back into Curtiss-Wright, and recovery from its post-war abyss seemed certain. However, it was sometimes said that, "the Air Force giveth and the Air Force taketh away."

When the USAF decided that the Northrop XP-89 Scorpion would be the next all-weather night fighter, the F-87A order was canceled on 10 October 1948.

The loss of the F-87 contract devastated Curtiss-Wright. With no other potential contracts in sight, Curtiss shut down its aircraft division and North American Aviation bought up all its assets. It was the end of an amazing and historically significant company in the annals of U.S. aircraft manufacturing. This career trajectory would be repeated by many of the existing aircraft manufacturers during the years to follow.

99

Chapter 3 — Secret US Proposals of the Cold War

A similar Curtiss-Wright XP-87 Blackhawk model was altered to display this nose turret proposal of the Glenn L. Martin 20-mm cannon. (Allen Hess collection)

Northrop F-89

Northrop was more successful in introducing their all-weather F-89 Scorpion. Although it was a subsonic aircraft fitted with straight wings, the F-89 still established a new precedent for manned Air Force interceptors in that it was the first aircraft designed specifically for the Cold War, able to carry out an interception day or night, in all conditions. Moreover, it was flown by a two-man crew – a pilot in the front seat, and a dedicated radar and weapon systems operator in the rear. By 1951, although the F-89 was fully operational with the USAF, Northrop realized that it wouldn't be long before their aircraft became outdated. The company proposed a new variant, with more fuel capacity and more powerful engines, called the "Advanced F-89." It looked good but other developments were already overtaking it.

Northrop's YF-89F all-weather fighter is represented by this wood Al Parker model. Originally Northrop's F-89 knocked the Curtiss-Wright XP-87 Blackhawk out of the competition, but when the F-89 started to show its age, Northrop proposed this YF-89F, their "Advanced F-89". Upon close examination of the fuselage one can see the family resemblance to the older F-89 Scorpion. (George Cox)

The mid-wing nacelles of the YF-89F each housed fuel tanks, the main landing gear, and in the forward section, fifty-two small air-to-air folding fin rockets. The aircraft was to be powered by two General Electric XJ73-GE-5 engines. Despite being an improvement over the F-89D, it was cancelled in August 1952. (George Cox)

The First Jets   Chapter 3

Grumman's Unknown Early Designs

During the jet age, Grumman, like any other aircraft manufacturer, continued developing new designs. And as new designs go, they are either refined or set aside depending upon their potential. Among the model collection at the Northrop Grumman Historical Center in Bethpage, Long Island, a number of unusual designs were found with a single identification tag marked, "Unknown."

As discussed earlier in this book, models have a variety of uses. They are made simply to confirm an engineer's 2D drawings, to see what a concept looks like in 3 dimensions, or for wind tunnel evaluation to refine flight characteristics. Ultimately they are to entice the customer and secure a contract. These Grumman "unknowns" appear to be preliminary concepts models.

Above: *This unknown Grumman wood model of an early jet design looks to be similar in form to the Curtiss XP-87 or the Gloster Meteor.* (Northrop Grumman History Center)

1st right: *One of the more odd designs found at the History Center is this wood model of an early jet with unusual forward and back-swept wings. The wing sweeps are separated by axial flow jet engines at mid wing.* (Northrop Grumman History Center)

2nd right: *Close-up of the second unknown Grumman wood model showing a splitter plate in the front of the engine. The solid Plexiglas piece forming the canopy is typical of early Grumman models.* (Northrop Grumman History Center)

3rd right: *A third in-house unknown Grumman with a high degree of wing sweep. Rather unusual in this design is how far back the engines are placed under the wings.* (Northrop Grumman History Center)

4th right: *Rear-three-quarters view of the same model showing to better advantage the unusual rear mounted engines off the trailing edges.* (Northrop Grumman History Center)

5th right: *The underside of this Grumman model showing the placement of the engines that are practically at mid-wing. Note also the air intakes set at an unusual angle.* (Northrop Grumman History Center)

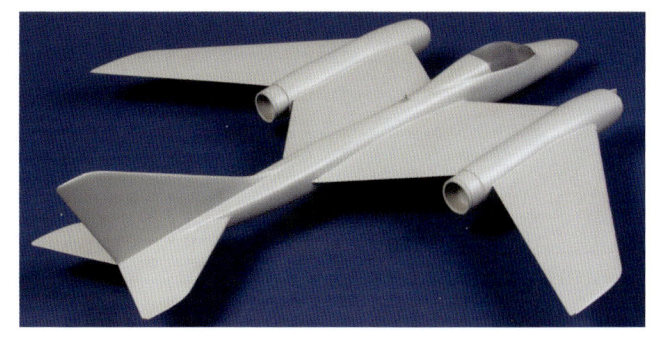

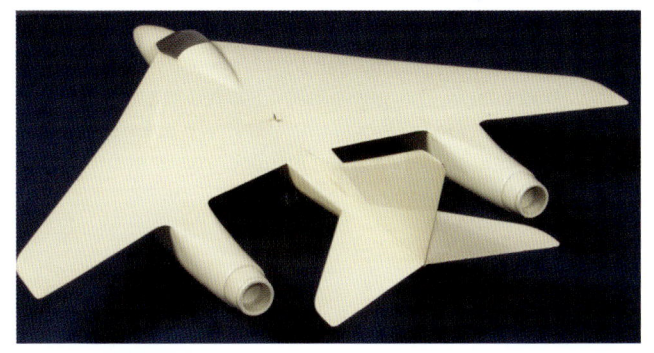

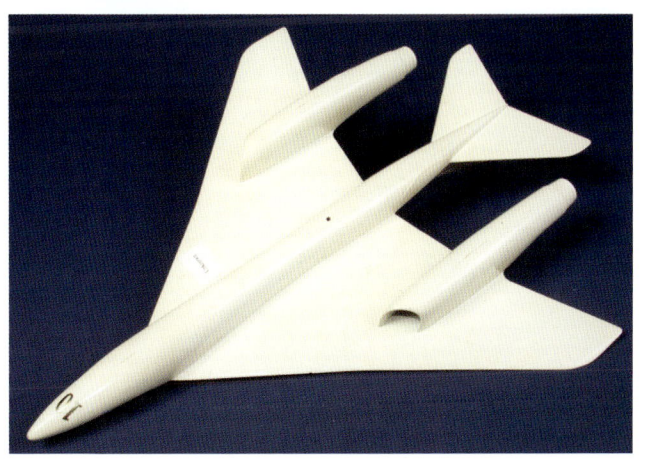

These unknown models show a diverse array of aircraft designs with their odd wing/engine configurations. They represent some of the thinking and exploration that went on at a time when the jet aircraft industry was just starting out. A few more of these Grumman "unknowns" will be presented later and in the Navy chapter.

Republic AP-31 The Thunderceptor

In 1945 the USAAF Air Material Command (AMC) had called for a very high-speed, high-altitude, daytime interceptor. This program came under Secret Project Number MX-809 (Materiel Experimental) and in December 1945, three months after the end of World War II, the AMC sent out an Invitation to Bid (ITB) to all aircraft manufacturers. The ITB included a Specific Operational Requirement (SOR) that stated the required abilities the aircraft must achieve. This interceptor had to climb to 47,000ft (14,325.6m) in 2.5 minutes, a rate-of-climb equal to 19,000ft (5791m) per minute. It had to cruise for 15 minutes at 486 knots (900.6km/h) indicated air speed (KIAS) and be able to sustain three minutes of combat speed at 688 knots (1,274.9km/h). Also stipulated was the ability to descend from 47,500ft (14,478m) and land in five minutes.

No doubt that the impressive German defence interceptor, the Me 163 Komet, was in the back of their minds when the U.S. military started thinking about their own advanced interceptor. But surpassing the performance of the Komet would require an aircraft that could approach the speed of sound. Yet no plane had ever flown faster. In fact, it would not be until 1947 when Captain Chuck Yeager Yeager would accomplish that feat—and he did it in a bullet-shaped, rocket-powered craft that was designed for only one purpose, to break the sound barrier. Add to this equation the fact that Yeager's aircraft could not carry armament, was not designed to take off under its own power, had only a few minutes powered flight before turning into a high-speed glider, and one can see the difficulties of approaching supersonic flight in 1945.

The military was pushing the aircraft industry beyond its capabilities. Maybe it assumed that aiming high would bring better results. The USAAF had set the bar high expecting a marked improvement in aircraft performance, but realistically, would have to settle with an airplane that came anywhere close to the SOR.

In January 1946, Republic responded to the call with the AP-31, (Army Project) a combined turbojet/rocket-powered interceptor proposal. On 29 May 1946 they received a contract to produce two prototypes. The Air Force designated them the XP-91, (later changed to XF-91) and it was named the Thunderceptor. This particular dual-power proposal was a first for a U.S. fighter and the rocket installation made it the first and only American rocket interceptor. The prototype XP-91 was built with an open nose air intake and a conventional tail arrangement. What was not conventional were the various features of the swept wings that had never been seen on a jet before. Most unique were the inverse taper and inverse thickness. The wings were wider and thicker at the wing tips than at the roots, where the wings joined the fuselage.

Republic model shop photograph shows a wood model variant of the AP-31 design. This solid nose model has divided air scoops on either side of the radome. There is a bulge between the tail and jet exhaust indicating an upper rocket housing. (Cradle of Aviation Museum)

Early AP-31 wood model showing yet another variation of the proposal. Note that this model has no radome, the nose has been opened for the air inlet. The unusual inverse tapered wings are shown with fuel tip tanks attached and has the V-tail configuration. (Cradle of Aviation Museum)

Now in its final configuration as a prototype, this in-house wood model of the Republic XF-91 exhibits the final air inlet as a large opening in the nose. This angle provides a better view of the unusual inverse tapered wings. (Cradle of Aviation Museum)

This inverse thickness provided greater outboard lift and minimized the onset of wing tip stalls, a bad characteristic in early swept-wing jets at slow speeds. Added to the wings were full-span leading-edge slats allowing the XP-91 to fly slower than any other jet fighter. The wing had variable incidence, ranging from -2 to +6 degrees, adjustable during flight for best angle of attack in take-off, landing, or cruise.

The proposed dual-power for the XP-91 consisted of the General Electric TG-190 series axial-flow J47 turbojet engine with afterburner, and a four-chambered Curtiss-Wright XLR27-CW-1 rocket engine. But by first flight, 9 May 1949, at Rogers Dry Lake at Muroc, the rocket engines were not ready, so the XF-91 flew on its turbojet and preformed quite well. Development problems forced Curtiss-Wright to cancel their rocket engines and Republic substituted the smaller Reaction Motors XLR11-RM-9 rockets, the same type that powered the Bell X-1.

On 9 December 1952, the XF-91 flew for the first time with the dual-power plants. Leveling off at 35,000ft (10,668m), test pilot Russell M. "Rusty" Roth lit up the afterburner for maximum boost, then switched on the rocket engines. The Thunderceptor flew past the speed of sound reaching Mach 1.07. This made the XF-91 the first jet airplane to break the sound barrier in level flight.

During its test phase, Republic experimented with different airframe arrangements. The first XF-91 was refitted with a chin scoop and a bullet-shaped nose that could house either rockets or the APS-6 radar system similar to the North American F-86D.

On request by the military the second prototype was retrofitted with the V-tail stabilizers and proved the engineers' predictions of drag reduction, increase speed as well as improved resistance to stall. With its dual-power plant it could climb to 40,000 ft (12191.9m) in one minute and had a top speed of 984mph (1,584 km/h) Both prototypes completed 192 test flights over the course of five years. Despite these achievements, the XF-91 was passed by. This was an aircraft designed at the end of the war, and by the 1950's, improvements of the turbojet made rocket power obsolete. And although the XF-91 never attained production status, the data collected from its design and construction was used to develop a more successful Republic jet, the F-105 Thunderchief.

Chapter 3 Secret US Proposals of the Cold War

A side view of the XP-91 in-house wood model, showing the conventional tail group as well as the outline of the rear housing for the afterburner and rocket motors. (Cradle of Aviation Museum)

Bottom: *F-91A tail close-up reveals a single rocket (possibly the proposed throttleable version) set further forward from the jet exhaust. Development of more powerful turbojets eventually eliminated the need for rockets.(Cradle of Aviation Museum)*

Below right: *Top view of the Republic F-91A production model with the radome nose, showing the unique inverse wings. This model incorporates the all-flying horizontal stabilizers and no longer has the leading edge spike ahead of the vertical. (Cradle of Aviation Museum)*

Below: *Identified as the F-91A on its stand and nose, makes this the production model that never came to be. It demonstrates the Thunderceptor with the nose radome, chin inlet and a single rocket in the tail. More than likely made to encourage the Air Force to order the XF-91 into production. (Cradle of Aviation Museum)*

Above: *The original tail configuration for the four chambered Curtiss-Wright XLR27 rockets of the Republic XP-91 that were never produced. (Cradle of Aviation Museum)*

The First Jets   Chapter 3

Again, another proposed production model of the F-91A, this time with the variations of the V-tail and solid radome nose. It is the companion piece to the conventional tail F-91A shown on the previous page. (Cradle of Aviation Museum)

Thinking that the military might want solid nose jets, Republic offered this second variant of the F-91A presentation model with a radome nose and NACA-style air inlets on the sides. (Cradle of Aviation Museum)

A rear ¾ view of the F-91A with the V-tail configuration. Republic's engineers had determined that the V-tail would provide less drag, increased speed and a reduced tendency of stalling. (Cradle of Aviation Museum)

Rear view of the proposed production F-91A V-tail model showing a slightly different arrangement for the Curtiss-Wright XLR27 rocket engine outlets. Compare this view with the rocket configuration for the conventional tail group shown on the previous page. (Cradle of Aviation Museum)

The Unknown Republic Proposal

Occasionally a model turns up representing a design for which no documentation appears to have survived. One interesting example is at the Cradle of Aviation Museum for which only speculation can be offered as to its nature. It represents a single-seat aircraft with small, stubby wings and a long empennage, probably holding an enormous engine. Like the Bell X-1, the canopy conforms to the sleek fuselage lines and the landing gear is contained in the under part of the fuselage. It is interesting to note the air intake located above the fuselage. All these features indicate an aircraft that was meant to go fast. This could possibly be another study for testing supersonic flight or a different concept for an interceptor.

The side view of this unknown Republic model displays the sleek lines to this aircraft. It also shows how thin the wings are compared to the horizontal stabilizer. Notice that the horizontal stabilizer has a spike on the leading edge that is ahead of the vertical stabilizer. This same configuration was used in the XP-91. (Cradle of Aviation Museum)

The positioning of the inlet scoop at the top of the fuselage presents an interesting arrangement and looks similar to the Lockheed L-205 or the later North American F-107. Notice that the wings are almost the same length as the horizontal stabilizer. (Cradle of Aviation Museum)

America's First Delta-Wing Aircraft

Convair's entry into the interceptor competition started as a swept-wing, dual-powered turbojet/multi-rockets, V-tail aircraft, similar to Republic's offering. Upon further development this proposal became a problem for Convair but its evolution would lead the company to a remarkable solution, the first delta wing jet.

On 12 April 1946, Convair was named winner of the interceptor competition, received a contract, given the designation of XP-92 and assigned MX-813 to the overall requirement. But wind tunnel tests showed

Conceptual art of the Convair Swept-Wing Interceptor illustrates a very sleek, straight forward design with a V-tail. It won the 1945 interceptor competition but would eventually evolve into the delta-wing XF-92A. (San Diego Air and Space Museum)

Another view of the Convair Swept-Wing Interceptor displaying 35-degree swept wings, V-tail configuration and rocket engines in the tail. This proposal was similar to Republic's AP-31 offering that also had a dual-power system as well as a V-tail variant. (San Diego Air and Space Museum)

This in-house model of the Convair XP-92 Interceptor is wooden with a plastic stand. The full-size fuselage was an 80-inch diameter metal tube with a mid-body 60-degree delta wing and an all-moving vertical fin. Main power was to be provided by a ducted ramjet with liquid fuel rockets externally mounted to boost take-off and climb. A Westinghouse turbojet was included for cruising and landing. (San Diego Air and Space Museum)

problems with lateral control and tip stalling. While researching the solution, the engineers came upon the German wartime papers of Alexander Lippisch's delta-wing studies. Adding delta wings to their design improved the performance and started Convair's long and successful hallmark, delta wing aircraft designs.

The XP-92 with its new wings took on a completely different look, something akin to a Buck Rogers spaceship. But then, this second design started to suffer from the complexity of its dual power-plants, causing delays in development and raising cost factors. Finally by 1947 the design was deemed incapable of reaching supersonic speeds and the military cancelled it as an interceptor. But they were still interested in developing a delta-wing jet. In the meantime, Convair tried working up other iterations to keep the program going.

The XP-92 pilot would fly in a prone position, housed in a large conical center-body spike that was partially buried inside the fuselage/engine housing. The intake airstream would flow over the spike and in case of emergency, the forward fuselage could detach, decelerate via parachute, allowing the pilot to escape. (San Diego Air and Space Museum)

In-house model of the Convair XF-92A represents the final evolution of the XP-92. Compare this model with the two earlier XP-92 proposals and one can see how the first airframes were combined to make up this final version. (San Diego Air and Space Museum)

To give more power to the XF-92A an afterburner was added. To try and save money, off-the-shelf parts were used; nose gear from a Bell P-63 King Cobra, main gear from a North American FJ-1 Fury, the ejection seat and cockpit from the XF-81, and the J33 jet engine that powered the Lockheed F-80 Shooting Star. (Barry Webb model)

A third proposal with a conventional fuselage, and a single turbojet was offered in September 1946. To keep the costs down, existing parts from other aircraft would be used. An order was placed November 1946 and it was designated the XF-92A. Two maiden flights occurred 9 June 1948 (a short hop) and 18 September (actual flight). The XF-92A became the world's first jet-powered delta-wing aircraft to fly, but still too slow as an interceptor. The military kept it on as a research aircraft and it was the first of Convair's delta-wing designs including the XF2Y-1 Sea Dart, XFY-1 Pogo, B-58 Hustler, F-102 Delta Dagger and F-106 Delta Dart interceptors.

A New Way of Thinking Interceptors

Relations between the Soviet Union and the West began to deteriorate significantly in June 1948, when the Soviet Union cut off all surface traffic to and from the Western-held sections of Berlin. It was a Soviet attempt to control Berlin and force the French, British, and Americans out of their sectors. The Truman Administration responded to this blockade with a daily airlift of food and supplies into the Western-held sections of Berlin and the famed Berlin Airlift lasted until the end of September 1949.

The crisis generated by this action escalated the ill feelings the U.S. military had for its former ally. With the anticipation of newer and more sophisticated Russian bombers coming in the near future, in 1949 the USAF opened a new competition for advanced interceptors. Since it would take about five years for U.S. aircraft manufacturers to bring these new interceptors into service, the competition was officially termed the "1954 Interceptor."

This new request called for an entirely different way of thinking and designing interceptors. Rather than just constructing an airplane, then attaching weapons to go seek a target, this new concept called for a complete "weapons system." From the start, the airframe would be integrated with radar-guidance, a fire control system, and selected aerial weaponry in the form of internally carried, heat-seeking or radar-guided air-to-air missiles. By 1950, these new requirements were formed into MX-1554 and called for a single-seat interceptor to carry air-to-air missiles, sustaining speeds in excess of Mach 1 and at altitudes beyond 50,000ft (15,240m).

The request was officially sent out in June 1950 and by the close of bidding in January 1951, nine projects had been submitted.

Lockheed L-205

Lockheed submitted a design that appears to be a precursor of the F-104 Starfighter, which actually came along a few years later. The model was listed as the L-205 and it had the similar characteristic of the F-104's small, thin straight wings (but mounted in the lower half of the body), the delta-shaped horizontal stabilizer, the long nose with the contoured canopy, and tricycle landing gear. This particular design had an unusual dorsal intake just behind the canopy, similar to the unknown Republic model mentioned earlier and the North American F-107 that came out a few years later. The L-205 carried six Hughes Falcon air-to-air missiles internally in a lower mid-fuselage

This view of the Lockheed L-205 in-house wood model provides a better look at the overhead air intake. A rarity among presentation models is the small pilot figure included in the cockpit. From this view, one can start to see the lines of the Lockheed F-104 starting to emerge. (Author)

Proposal for an interceptor, the Lockheed L-205 has quite a thick fuselage. It needed the space to internally accommodate six Hughes Falcon air-to-air missiles, turbojet, and main landing gear. Note the engine spike coming out the tail exhaust. (Author)

section along with twenty 2.75in (6.98cm) folding fin rockets housed in the sides of the bay.

Lockheed provided the following estimated performance specifications:

Time to climb 40,000ft (12,192m)		1.6 minutes
Combat ceiling	63,000ft	(19,202m)
Maximum range	1,530nm	(2835km)

Republic's AP-54, AP-57

Republic came up with three different proposals to enter into the "Interceptor 1954" competition. Each offered three different types of power-plants, for Republic was trying to cover all its bases in hopes that any one of the projects would be selected. One proposal had a dual-power configuration similar to the XF-91 with a turbojet and a single rocket combination. The second simply had a large turbojet with afterburner. And the third also came with a dual-power system, but with a different combination consisting of a turbojet and a ramjet that shared the same intake and exhaust. This last entry, the AP-57, was a stunning design. Among other things it was created to explore the possibilities of ram jet propulsion to get beyond Mach 3. For the early 1950s this was definitely a jump into the future. Republic's AP-54 was a more advanced design of its predecessor, the AP-31/XF-91, "AP" now stands for "Advanced Proposal." It is included here for it falls within the time period and, although not verified, the AP-54 may well have been one of Republic's three designs (the turbojet and single rocket engine configuration) entered into the "Interceptor 1954" competition. It has a sleeker fuselage, inverse wings and horizontal tails with more degrees of sweep, and blended, thinner wingtip tanks. Just looking at the model, it is already attaining Mach 2 while standing still. If the AP-54 and the AP-57 were actually offered for the interceptor competition, then one could speculate that Republic was playing both sides of the technological street. They were offering an advanced version of a tried and tested design of the XF-91, while at the same time they were dangling an amazing, revolutionary technological leap into the future – the AP-57. And it caught the eye of the military.

Republic's In-house wood model of the AP-57 portrays an aircraft of advanced design for the early 1950s. It appears to be more missile than interceptor, partly because the canopy was eliminated in order to reduce drag during supersonic flight. Max speed was calculated at 3.7 Mach. (Cradle of Aviation Museum)

The Republic AP-57/XF-103 interceptor is now shown with its ordnance extended from retractable housings. The XF-103 was intended to carry six MX-904 (GAR-1) Falcon missiles and 36 2.75-inch FFAR rockets . (Cradle of Aviation Museum)

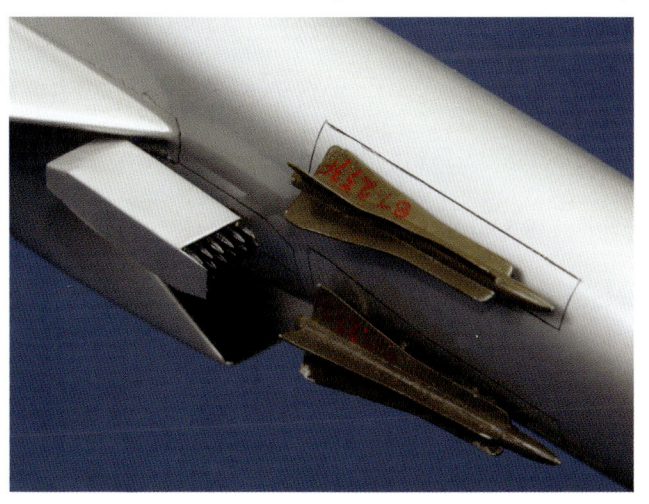

Top view of the Republic AP-54 shows Republic's styling of swept-back inverse tapered wings. (Cradle of Aviation Museum)

The in-house wood model of the Republic AP-54 with its slim wingtip tanks helps give it that fast appearance. Note the rocket outlet beneath the jet exhaust. (Cradle of Aviation Museum)

Chapter 3 — Secret US Proposals of the Cold War

The Republic AP-57/XF-103 with escape capsule (not to scale) known as "the shoe." It would eject out the bottom of the fuselage, and be able to withstand speeds up to Mach 3.7 for pilot survivability. (Cradle of Aviation Museum)

On 2 July 1951, the USAF selected three designs from the "1954 Interceptor" competition for further development, the Lockheed L-205, Convair's concept (eventually the YF-102) and Republic's AP-57/ XF-103. Comparing the model of the XF-103 with the AP-54, or even the L-205, one can see a marked difference in the look and feel between these proposals, the XF-103 has a much more powerful and futuristic sense.

Republic had shifted advanced design into a higher level by offering a number of "firsts" with their ambitious project. Republic proposed to fly the XF-103 beyond Mach 3, have a climb rate of 66,000ft (20116.8m) per minute, and reach a height of 80,000ft (24,384m). To counter the aero-thermodynamic heating caused by high speeds, the XF-103 would be the first aircraft to be constructed out of titanium. And the first to have a special blue paint coating given to the entire aircraft to help radiate the heat away from the airframe. To reduce drag the canopy was eliminated, substituting a flush glazing or window and the pilot would use a periscope for a forward view. Another first for the XF-103 would have a pilot's escape capsule for survivability at supersonic speeds. It was to carry internally six Hughes GAR-1 Falcon air-to-air missiles in separate bays along the forward fuselage sides.

To achieve the high Mach numbers Republic, once again, would build an airframe with a dual-cycle propulsion system consisting of a turbojet and ramjet combination. Curtiss-Wright would develop the J67-W3 turbojet with an afterburner that housed the ramjet. This unit would weigh 7200lb (3265.9kg) and take up half of the 81ft (24.7m) long fuselage. Republic was ordered to build three prototypes of the XF-103 and the USAF now termed this aircraft Weapon Systems WS-204A.

Ramjets were of great interest to the military in the early 1950s because of the turbojets limitation which at that time

The First Jets

could only reach speeds under Mach 2. Ramjets were the only air-breathing propulsion system with the potential to power an aircraft into higher Mach numbers. But ramjets need a fast inflow of air before they can function, so the turbojet would get the airframe flying fast enough to give the ramjet the air it needed to kick in. Once up to speed, ducting allowed air to by-pass the turbojet and flow into the ramjet, which then became the interceptor's power source.

But the XF-103 was six years in development, at a cost of $100 million, and it was still a year or two from completion. And in the same scenario as with the Thunderceptor, Curtiss-Wright could not deliver the dual-cycle engine with the promised thrust. The project had gone on too long, and military's interest now turned to long range interceptors, In August 1957 the XF-103 was cancelled.

Grumman's Unknown Ramjet Design

This unknown Grumman model is presented here because it fits into the topic of ramjet propulsion. The design and age of the model appears to be from the early 1950s, and the canards and tail are comparable to the XF-103 in looks. The model may be a study for a dual-cycle turbojet/ramjet or possibly multi-ramjets. The air-inlet scoops around the mid-section of the fuselage are typical for a ramjet, while the large extended rear opening could be for a turbojet afterburner. There is no canopy painted on this model, which might just make it simply a rocket proposal, but the bulge in the nose of the model would indicate otherwise. Here again is a model that could be considered a simple design study but also a puzzle piece from Grumman's history requiring more investigation.

Higher and Faster

Multiple, overlapping mission requirements now began to drive these entirely new breeds of aeronautical development. During the 1950s, the distinct trend in U.S. interceptor and fighter development was for ever-larger, heavier, and more complicated aircraft. This also involved more complex radar and fire control systems to manage the weapons. After the experiences of fighter combat during the Korean Conflict, however, another new trend began to emerge; one that called for smaller, lighter, and faster fighter aircraft. The wish list now included a fighter that was a pure jet, light, agile, simple to operate, and ideal for dogfighting. The Korean experience also showcased the need for fighters to operate out of forward airbases, climb quickly to combat altitude, and fight efficiently at these high altitudes.

Ramjets were explored as a solution to attain Mach 3 flight. This unknown Grumman design has the inlets that were typical for a ramjet engine. Not typical of Grumman are the canards. (Larry McLaughlin collection)

A Grumman in-house wood model of an unknown ramjet design. Again, speculation offers an explanation that the bulge in the nose may be the makings of a canopy that was not completely painted in. (Larry McLaughlin collection)

The next call for faster and better aircraft raised the bar, again, on aeronautical design, requiring top speeds of Mach 1.3 at 35,000ft (10,668m) with a 350nm (648.6km) radius of action. It also required aircraft to fly from runways as short as 3,500ft (1,067m) and to be armed with at least two 30mm cannons or the "equivalent firepower" capable of executing four two-second firing passes per mission. Hence, the "Interceptor 1954" competition was not even concluded when, in 1953, the search began for even faster, more agile, and more technologically advanced fighters.

North American Supersonic Lightweight Fighter Rapier III

At this time, little is known about the model of the North American Supersonic Lightweight Fighter Rapier III. It is placed in this section because of its size and judging from its overall design, does seem to fit the time period. But this is only a guess. Again, the quality of this model is superb; the high-gloss finish and the nicely executed canopy give it an appealing look. North American was one of the leading makers of high-quality models, particularly from the 1950s and 1960s..

Republic AP-55

Very few models have appeared with inverted V-tail arrangements. A Republic concept was one of the few found with this unusual tail. Republic experimented with its V-tail or "butterfly"-style tail on its second XF-91 as well as it on its NP-series, for the Navy (see Chapter Four). Two different versions of the AP-55 were examined. Close observation shows that the canopies are different, as are the fuselage lengths, air inlets, and tail cones. Republic did a number of variations of the AP-55, which seems to indicate that the company spent quite a bit of time investigating the potential of this aircraft.

Viewing the North American Rapier III model from this angle reveals an appearance similar to the F-100 Super Sabre. (Author)

This in-house wood model, represents a lightweight supersonic fighter, the North American Rapier III, but little is known about it. The Rapier III may to have come after the F-100 and F-107. (Author)

The First Jets   Chapter 3

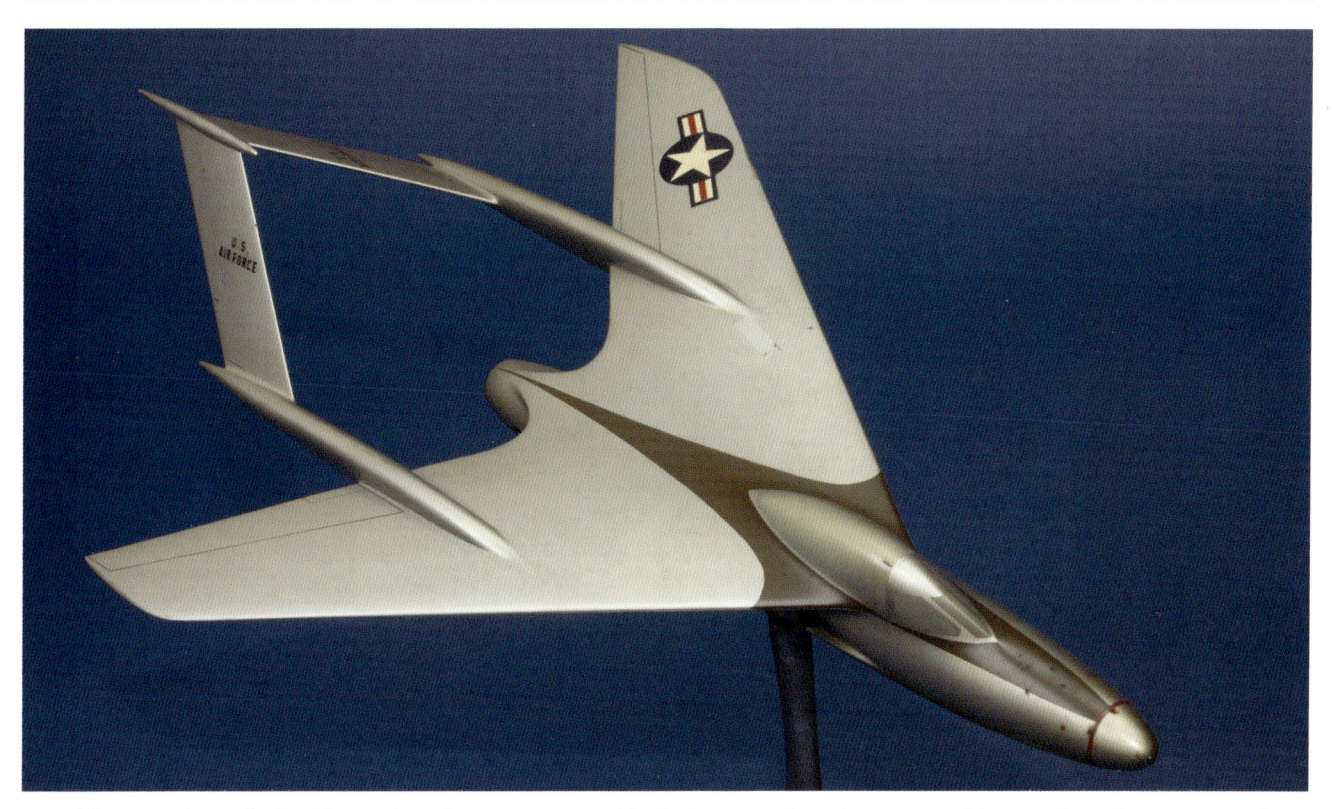

Republic not only worked with V-tail configurations, as with their XF-91, they also proposed inverted V-tails as seen on this model of their AP-55 proposal. (Cradle of Aviation Museum)

Here is a different pose of V-tailed Republic AP-55's in-house wood model. Notice the NACA-style inlet underneath the wing on the side of the fuselage. (Cradle of Aviation Museum)

A closer view of the elongated AP-55 variant, also an in-house wood model, showing a thinner fuselage with triangular style air inlets that extend beyond the leading edges of the wings. (Larry McLaughlin collection)

Republic AP-55 variant with a longer fuselage and completely different inlets placed ahead of the leading edge of the wings. Also note that the canopy is fared into the fuselage. (Larry McLaughlin collection)

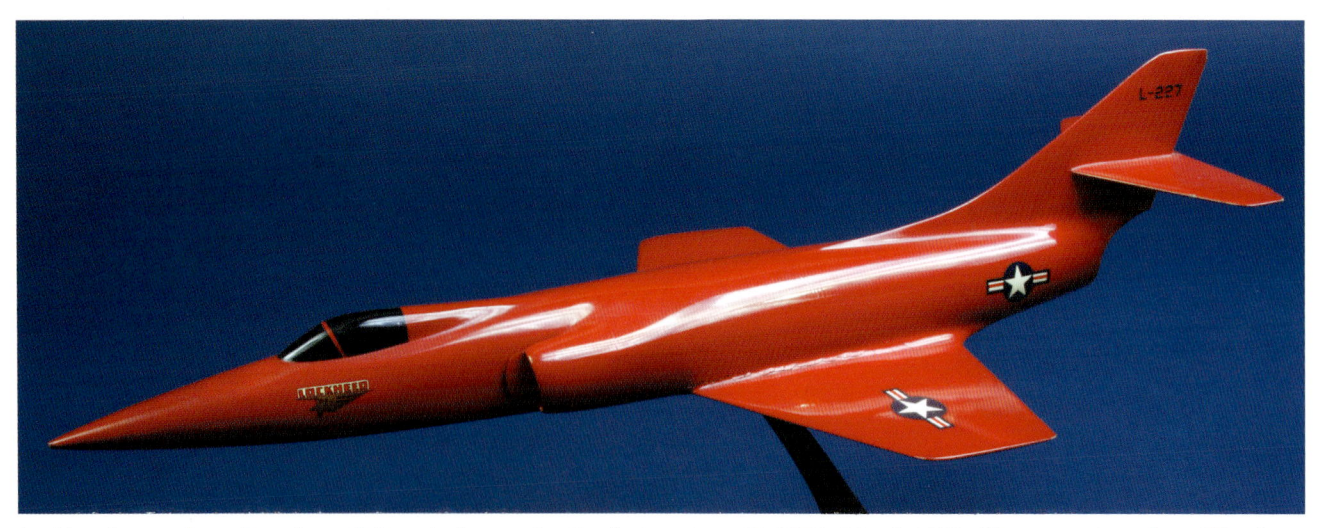

Lockheed started a series of new fighter designs under the designation of L-227 in March 1952. After a number of modifications on paper, the resulting model was this L-227-1, an early iteration of the F-104. (Author)

This in-house wood model with a fire red lacquer finish made Lockheed's L-227-1 stand out, considering that practically every other model from this era was standard silver color. One collector dubbed it the "Ferrari" of Cold War presentation models. (Author)

Lockheed L-227-1

Compare the Lockheed L-227-1 model to the Lockheed L-205, and it appears that the L-227-1 evolution leads, eventually, to Lockheed's famous F-104 Starfighter. The nose and side intakes, as well as the short wings, start to show the familial lines. In this case, the tail group arrangement of the L-227-1 has not reached the form of a T-tail. The all-red paint is very unusual for factory display models and with the gloss-black canopy, this aircraft looks more like a race plane than a military one. It is a very impressive presentation.

Search for a Long-Range Interceptor

The concept of a dedicated interceptor was first applied in earnest by the Germans during the closing months of World War II, when they could not defend against the onslaught of Allied bombers that flew day and night over their country. The Germans developed the remarkable Messerschmitt Me-163: an aircraft that could rocket up to 40,000ft (12,192m) in about 4 minutes, level off, and then dive on B-17 and B-24 formations while firing 30mm cannons. The idea was simple: take off, climb quickly, intercept and shoot down enemy aircraft before they had a chance to drop their bombs. Given the aircraft's incredible performance, this might sound like a straightforward mission. However, the German pilots had three potential enemies to contend with—the Allied gunners in the heavily armed bombers, the P-51 escort fighters, and the volatile (and unstable) rocket-fuel chemicals that carried them aloft. The Me-163 could only carry a limited amount of fuel – enough for about 5 minutes powered flight. Once out of fuel, it became a glider coming down for a dead-stick landing. During this

The First Jets
Chapter 3

Above: *A close view of the Douglas Model 1355 forward fuselage and two Wright J67-W-1 engines. The stubby wings and large engines give it the appearance of great speed, a necessary quality for an interceptor.* (George Cox via Tony Buttler)

Right: *An overhead view of the Douglas Model 1355 shows to advantage the squared-off wingtips that were used as tip ailerons. This model is made from wood with a silver lacquer finish and engraved lines to show the panels and control surfaces.* (George Cox via Tony Buttler)

return to base they were highly vulnerable to the American fighters. And on landing there was a further danger. If there was any un-spent fuel in the tank, it could explode with the impact of touch-down!

During the subsequent Cold War, American interceptor pilots may have had a little more confidence in the technology that carried them into the sky, but now the stakes were much higher: a single bomber getting through could lay waste a city. The idea, therefore, was to create an aircraft that was not just a point defense interceptor, but a long-range interceptor; one that could locate, intercept and destroy enemy bombers long before they could reach the targeted area.

As early as 1952, the military considered possibilities for a Long-Range Interceptor (LRI). On 6 October 1955, the USAF called for an interceptor that could fly at Mach 1.7 at 60,000ft (18,288m) with a range of 1,150 miles (1,850.8km). Additionally, it should include powerful radar to seek out enemy bombers at a range of 60 nm (69 miles/111 km) and employ an integrated fire control system capable of destroying up to three targets during a single mission. These stipulations were designated by the military as Weapon Systems 202A (WS-202A). An invitation was issued to bid on this project and several aircraft manufacturers responded. Some of the designs to be submitted were as follows:

Douglas Model 1355

The Douglas Model 1355 was a two-seat proposal designed for speed with sweptback wings and a contoured fuselage to reduce drag and permit transonic and supersonic flight. The huge engine pods mounted with pylons under each wing were placed well forward of the leading edges. They contained two Wright J67-W-1s with afterburners and thrust reversers in the aft end. This design had an all-flying horizontal tail, single slotted wing flaps, and squared off wingtips that functioned as tip ailerons. To aid its interception mission, this aircraft used a Doppler navigation system and a combined all-weather search and fire control system.

Lockheed CL-288-1

The Lockheed CL-288-1 design was submitted in June 1954, and is a much larger version of the F-104 Starfighter. Although bigger, it used the same general configuration as the F-104. Directly taken from the F-104 was the same tricycle landing-gear design and downward-ejecting aircrew seats, while the wings, tail group, and fuselage were modified, larger versions of F-104 components. The engines were placed in mid-wing nacelles instead of incorporated within the body of the craft. Also, because it was a larger aircraft, the main landing gear featured double tandem wheels mounted on each side.

This 1954 in-house wood model of the Lockheed CL-288-1 represents a proposal for a larger version of the Lockheed F-104 to perform as a dedicated interceptor. (John Aldaz collection)

A close-up of nose and engines of the in-house Lockheed CL-288-1 wood model. The high-polished silver and red lacquer paint job makes it a very appealing model. (John Aldaz collection)

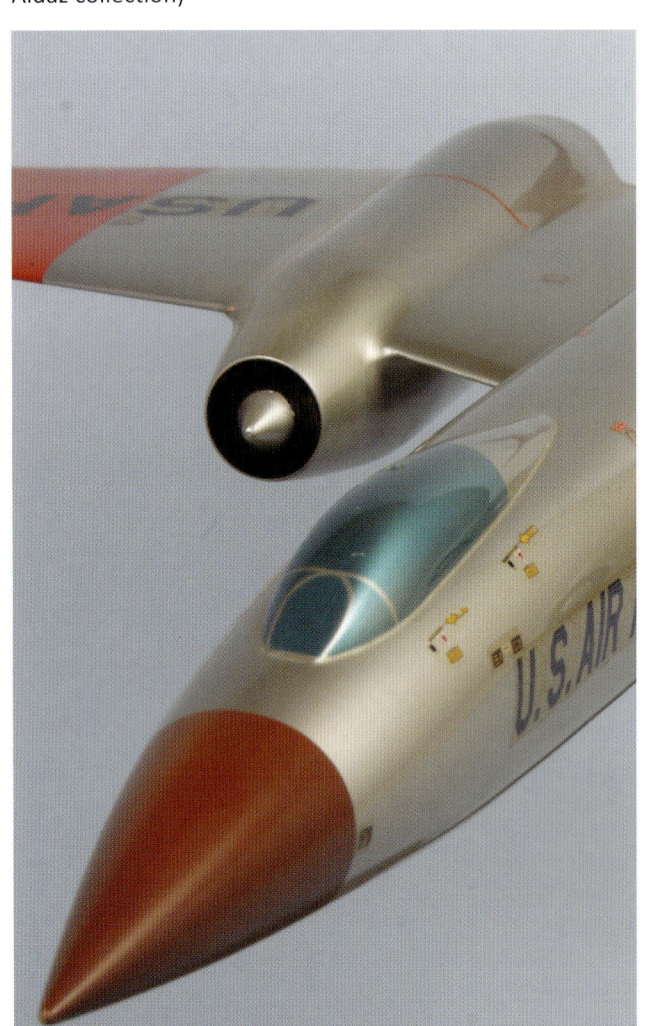

The model has a distinctive paint scheme. The outer wings and the tail group are painted bright red, which sets off the model's overall silver coloring. This seems similar to the bright Insignia Red-Orange used by aircraft operating in snowbound regions, but seems more likely to be a modeler's artistic license in coloring to give the model more visual appeal.

North American Long Range Interceptors

The first designs of the North American LRI model have a very futuristic appearance with much sleeker lines than many of the designs previously seen.

In model form, two variations of the same aircraft were encountered. One, slightly larger than the other with wing-tip tanks is 26in (66cm) long. It is speculated that this in-house study is North American's first attempt at the LRI, maybe as early as 1953/54. The stand labels it as an "Advanced Piloted Interceptor". The second model, without wing-tip tanks, is 18in (45.7cm) long, and is slightly different in wing form and canards. Its stand identifies this model as the WS-202A, probably from 1954/55.

A close-up shot of the nose of the Advanced Piloted Interceptor shows the airbrushed reflections detailing the canopy and the squared-off air intakes. This model was made in the early 1950s. (Author)

It is speculated that this is North American's early in-house study for a Long Range Interceptor. The model stand is inscribed "Advanced Piloted Interceptor". Compared to the later WS-202A, one can see the wing and canard shapes are different, the API has wingtip tanks and the fuselage has a slightly different form. (Author)

The next iteration of the Long Range Interceptor is represented by this in-house wood and metal model from North American, the WS-202A. North American was one of the best model makers of this period, not only for the quality of its models but also for its designs. Notice how the stand is incorporated into the back of the model with red paint covering the supports that emerge from the tail pipes as hot exhaust. (John Aldaz collection)

Top: *Official black-and-white company photograph showing the ESO7189 from a level angle. This particular model has an under-wing fuel tank. This was the final version North American submitted as their LRI.* (Author)

Above: *Another view of the North American ESO7189 model shows it without the under-wing fuel tank. This final version appears as a very conventional looking aircraft when compared with the previous proposals.* (Author)

These models definitely show the craftsmanship that went into producing these pieces of art. Made primarily of wood, the wings are constructed from laminated wood strips. This was done to keep the wings from warping. The canards and the tail fins are metal.

An interesting feature of the smaller model is the stand. The aircraft is mounted to the stand through its exhaust pipes. The stand is colored red at the point where it holds the model to simulate the heat coming out of the exhaust – a very nice touch by the model makers at North American.

Finally, there is a third model found in black and white photographs identified as the ESO7189LRI from North American files. According to documents found at the National Archives, this was the model that was submitted by North American as their Long Range Interceptor in 1955.

Grumman Design Number G-107-3

By June of 1954 Grumman had developed a proposal that met or even surpassed all the requirements for a new interceptor. The model it produced had a 45° sweep to the midsection wings that held twin Wright J67 engines with afterburners. The engines were configured in an unusual position: They were stacked one on top of the other, housed down the centerline of the fuselage. The air inlets were on either side and just behind the canopy. It was estimated that this aircraft could hit a top speed of Mach 2.14 and climb at an incredible rate of 48,500ft/min (14,782.8m). But the company chose to withhold its bid because the aircraft it came up with was too large to be built in its facility. The model shows a very large wing with a contoured fuselage for transonic and supersonic flight. The purpose of the twin booms on the wings is not known, but is speculated to be fuel tanks. The tail surfaces were also swept back, and the canopy made large enough to accommodate side-by-side seating of the pilot and copilot.

The Grumman Design Number G-107-3 in-house model shows the unusual configuration of twin Wright J67 engines stacked one above the other with afterburners. (Northrop Grumman History Center).

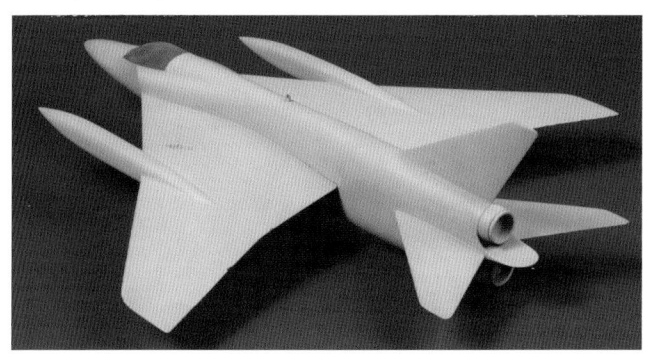

The First Jets Chapter 3

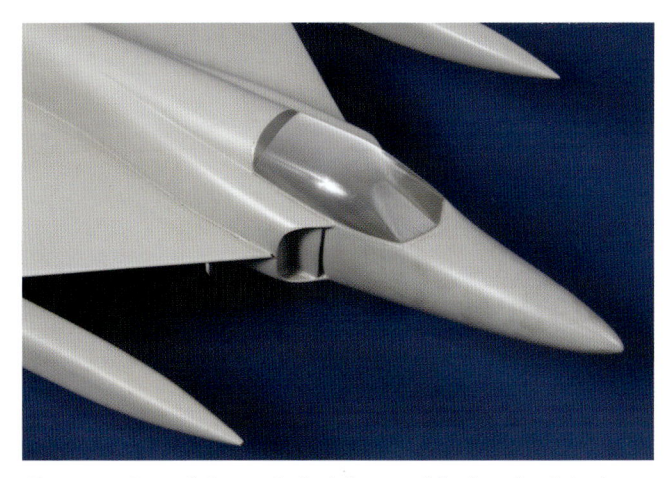

Close-up view of the cockpit, inlets and fuel tank of the in-house Grumman G-107-3 model. The wings of this aircraft have 45 degrees of sweep inboard of the fuel tanks and 35 degrees on the outboard section. (Northrop Grumman History Center).

The Ultimate Long-Range Interceptor

By the mid 1950s the competitors in the race to develop a long-range interceptor were evaluated, and all were found to have fallen short of the stipulated performance goals. But the Air Force decided to continue Long Range Interceptor development with Northrop, Lockheed, and North American. In October 1955, the USAF now called for the LRI-X, (long range interceptor, experimental) to be a two-seat, twin-engine aircraft, flyable by 1961.

Ultimately North American was chosen to develop what became known as the F-108 Rapier, the ultimate long-range interceptor. North American made several progressive alterations to its interceptor project to get to the final design.

The North American wood and metal in-house model of the NA-257 represents an early version of the XF-108. It has three vertical fins and navigator's windows just behind the cockpit. (Author)

North American NA-236 and NA-257 (F-108)

The road to developing a long range interceptor was a difficult one. In January 1956, the Air Force selected the NA-236 as the winner of the LRI competition, and in May the Pentagon canceled the program—only to reinstate it in April 1957. In June, North American received a contract to develop the aircraft, now designated by the manufacturer as NA-257 and by the military as F-108. This design went through many changes inside and out. The canards and upper fins on the trailing edges were removed, and the wing took on a 53.5° sweep angle. Eventually the wing was changed again to a "cranked arrow" shape and tip extensions were added.

The F-108 received the new General Electric J93-GE-3R engine (the same as those to be used in North American's larger XB-70 Valkyrie bomber), and the fire control systems were updated. In fact, the F-108 was to be a fully automated aircraft with the pilot only having to exercise manual control on takeoff and landing. Because of the expected Mach 3 speed at which this aircraft would fly, new groundbreaking materials and alloys were developed, as well as new fuels. All this research was also employed in the XB-70 that came through its development on the heels of the F-108.

A close-up of the nose and inlets of the North American NA-257 model shows the two side windows for the navigator reflected in the canard. Other versions eliminated these windows. (Author)

Chapter 3 — Secret US Proposals of the Cold War

A North American artist's concept painting of the XF-108 shows the aircraft in a dramatic pose, giving it the aura of a strong and powerful interceptor. Squared air intakes were quite novel for this time period, but would later appear on a host of modern U.S. and Russian fighters. (Tony Buttler Collection)

Another North American concept painting now shows the XF-108 with a large single vertical stabilizer replacing the twin and triple tailfins of earlier studies. (Tony Buttler Collection)

By November 1958, an operational date of mid 1963 had been set and the full-size mock-up was presented at the North American facility in January 1959. North American Aircraft had in recent years started a policy of naming its Air Force fighters after styles of swords, so in 1958 the F-108 was named the Rapier (following on from the F-86 Sabre).

The firepower capabilities of this aircraft were amazing for its time. It utilized the Hughes AN/ASG-18 fire control system to manage GAR-9 missiles. From an altitude of 70,000ft (21,336m), the F-108 could launch these missiles to seek out and bring down any three air-breathing targets flying at altitudes anywhere between sea level and 100,000ft (30,480m).

But the cost to develop this aircraft kept escalating and the USAF could not keep up with the funding. Eventually the military was forced to begin cutting back the scope of this technological beast. It was a slow downward spiral. In August 1959, the Rapier was put into the "strictest austerity" program and in September, the project was ordered to close.

Company photograph of North American's in-house wood and metal model of the XF-108. This elegant design shows the power and grace of their long range interceptor in a version that is very close to its final configuration. (Author)

Company photograph of the North American F-108 model displayed with its various support vehicles and ground equipment surrounding the aircraft. This is the final version of the interceptor had it been built. (Author)

Lockheed YF-12

Following the cancellation of the Rapier, the USAF had nothing in its arsenal to succeed its fleet of Mach 2 F-106 interceptors. At this point, Lockheed entered the picture. It had built a Mach 3 aircraft, a remarkable high-speed, high-altitude photo-reconnaissance for the CIA as a replacement for the U-2 spyplane. Lockheed's A-12 had first flown on 26 April 1962, and had already demonstrated the ability to fly at over Mach 3.2 and attain 90,000ft (27,432m).

Lockheed was able to offer a fighter version of the A-12, re-designated the YF-12A, with a Hughes radar guidance and fire control system to manage the Hughes AIM-47 air-to-air missiles. It became the largest, the heaviest, the fastest, and the highest flying manned interceptor ever flown.

Three prototypes were built that set several absolute speed and altitude records surpassing all expectations of what previous proposals had hoped to do. How ironic, that by the time the YF-12A was ready for delivery the Air Force never ordered them into production. Time and technology had changed the end game. By the early 1960s there was no longer a need for long-range interception of enemy bombers, for nuclear bombs could be more efficiently delivered by the ICBMs.

The YF-12 was cancelled in 1968 but another aircraft that had been in development went on to assume the original role of the Y-12 as a strategic-reconnaissance aircraft, the SR-71.

The Lockheed YF-12A first flew on 7 August 1963. This version was armed with the Hughes AIM-47 air-to-air missiles and powered by two Pratt & Whitney J58 engines. Italeri 1/72-scale model. (Barry Webb model)

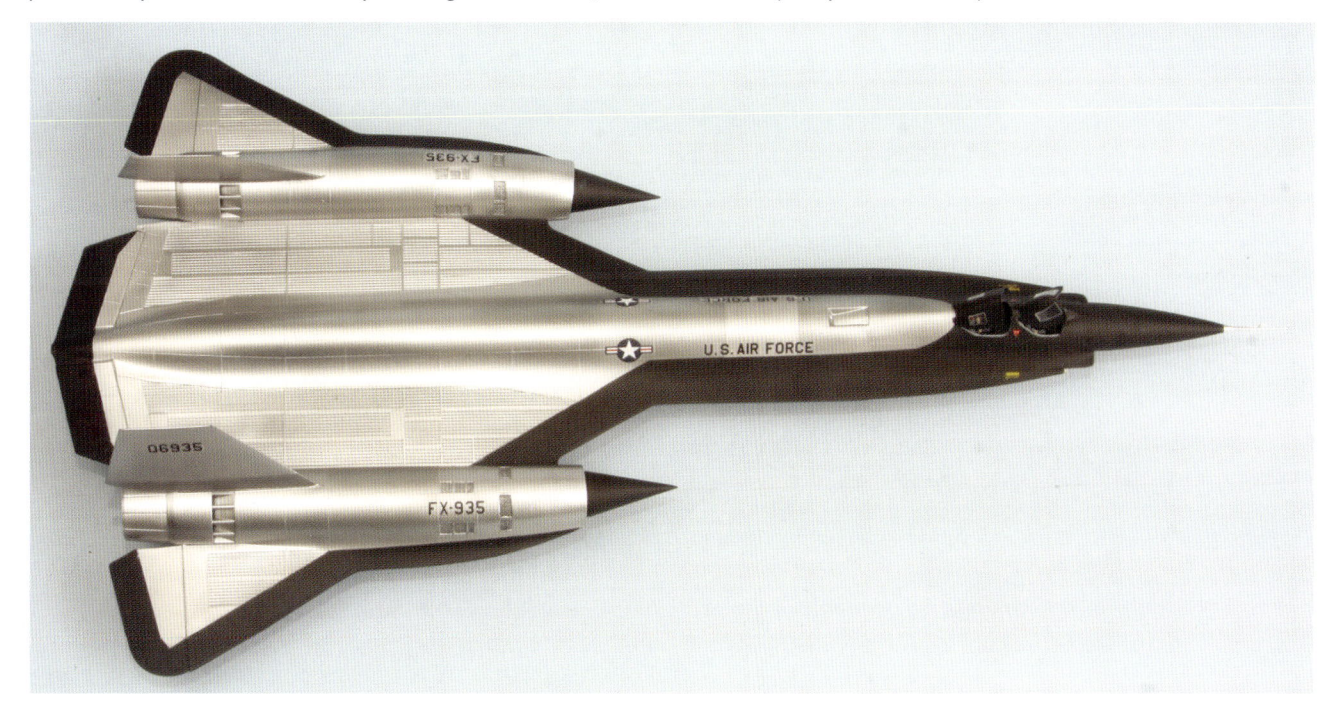

Chapter 3   Secret US Proposals of the Cold War

The Lockheed SR-71 Blackbird. This unarmed aircraft became the ultimate surveillance plane during the Cold War. Note the differences between this model and the YF-12A in the nose and the tail. Testors 1/48-scale model. (Allen Hess collection)

Cruver manufactured this ID model for US Forces representing the Soviet Ilyushin Il-54 bomber, known by the NATO codename 'Blowlamp'. It was issued in March 1958, when it was thought that this aircraft was about to appear in squadron service. In reality it had been cancelled two years earlier. *(George Cox)*

ID Models in the Cold War

During the Cold War, both sides developed new bombers and fighters at a rapid pace, and the need to teach aircraft recognition remained a preoccupation for the American and Allied forces. The ID models that had proved to be so useful during World War II were now back in high demand, and with so many new aircraft being produced, the manufacture of these small ID models continued unabated.

The First Jets Chapter 3

These two NATO models made by Verkuyl are to the same scale, 1/144. They illustrate the huge leap in size that the B-36 represented over the B-29, the latter being far and away the heaviest of WWII bombers. *(George Cox)*

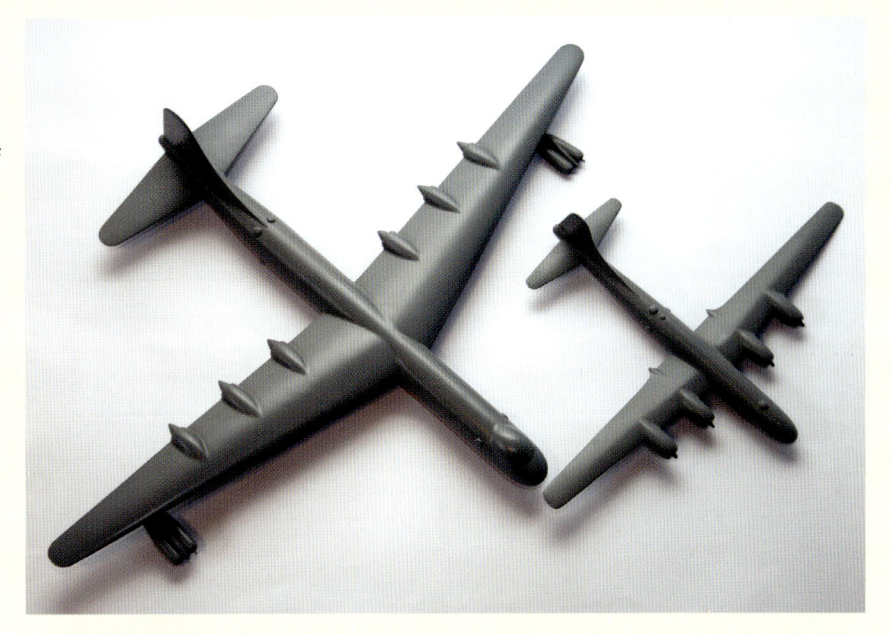

A Cruver 1/144-scale model depicts the very elegant Tupolev Tu-85 Soviet bomber given the NATO codename 'Barge'. It was issued in July 1955, by which time the sole prototype had long-since ceased to fly! *(George Cox)*

The United States continued with the Cruver plastic 1/72 series, but to help designate the difference from World War II to the Cold War the color was switched from black to gray (with a few models finished in silver and blue in the interim). The series was also supplemented with a range of smaller-scaled models; 1/144 to cover the larger airplanes.

The British continued producing 1/72 wooden models, also switching from black to blue-gray, and the Dutch, using the outstanding skills of model-maker Matthijs Verkuyl, produced aluminum models for the NATO forces in Europe, using both 1/144 and 1/72 scales.

The Germans, meantime, stuck to their preference for tiny 1/200 models, but after Wiking declined to make them, the Germans switched to Hansa as the supplier. The Hansa models were made of metal (for the smaller planes) and of plastic (for the larger ones). They came in double-sided wooden carrying cases, making them easier to transport and enabling instruction sessions to be given on complete fleets of potential friendly and enemy aircraft.

Throughout the Cold War, Soviet developments were shrouded in secrecy, and the ID models often reflected the lack of accurate information. For example, the Verkuyl model of the Mikoyan MiG-19 was nothing like the airplane that was actually to carry this designation; instead it was represented as the German World War II Focke-Wulf Ta-183 design, on which Western intelligence believed the Soviet plane was based!

Other models in the western ranges represented Soviet types assumed to be in squadron service but which, in fact, never advanced beyond the prototype stage. All-in-all, these ID models, many of which survived thankfully, represent a miniature history of aircraft development during the post-war years.

123

Chapter 3 — Secret US Proposals of the Cold War

The German NATO Forces used very small 1/200-scale models for identification training during the Cold War era. However, the wartime Wiking models were succeeded by miniature aircraft made by Hansa. They were issued in double-side dark green, lockable wooden boxes. The picture shows one such box, Set #4 containing the Soviet aircraft. The only one aircraft that is not quite to scale is the 'Bear'. *(George Cox)*

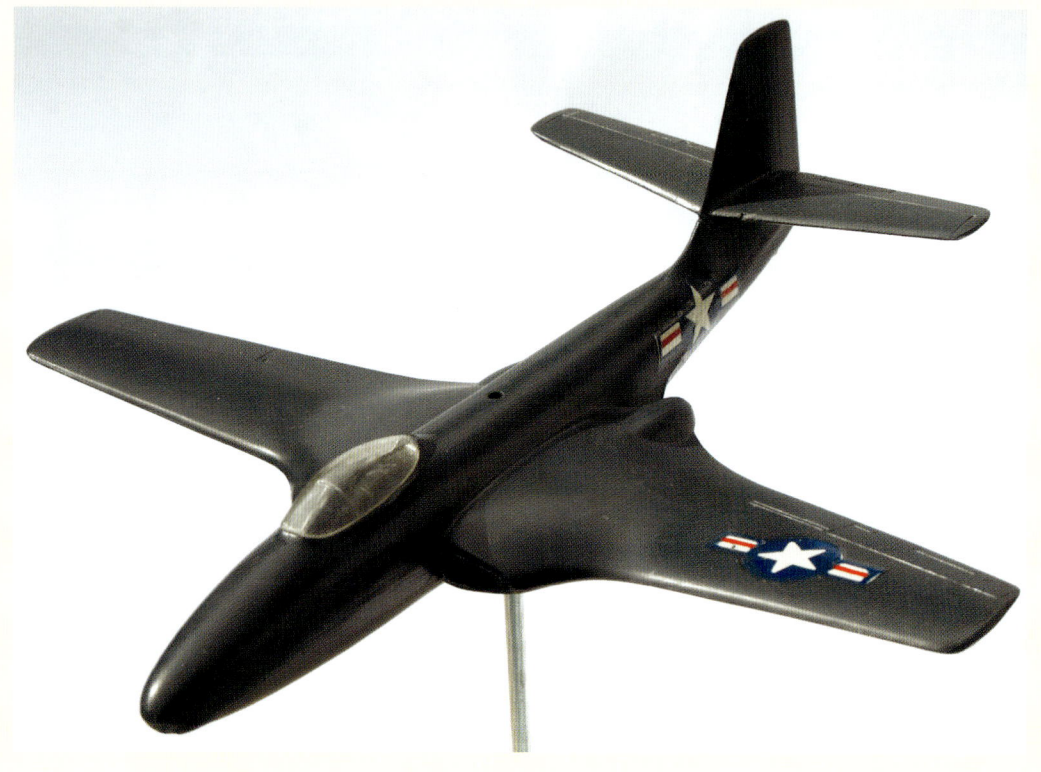

A good example of a typical post war ID model made for the US forces is this very neat little 1/72-scale representation of the McDonnell F2H Banshee issued in July 1950. It is finished in midnight blue, a color used for US Navy ID airplane models for a very short period. *(George Cox)*

The First Jets
Chapter 3

After the shocking appearance of the Soviet MiG-15 in Korea, there was much speculation about what would follow it. This NATO ID model made by Verkuyl of the so-called 'MiG-19' was erroneously based on the German Focke-Wulf Ta-183. Such a plane never flew. *(George Cox)*

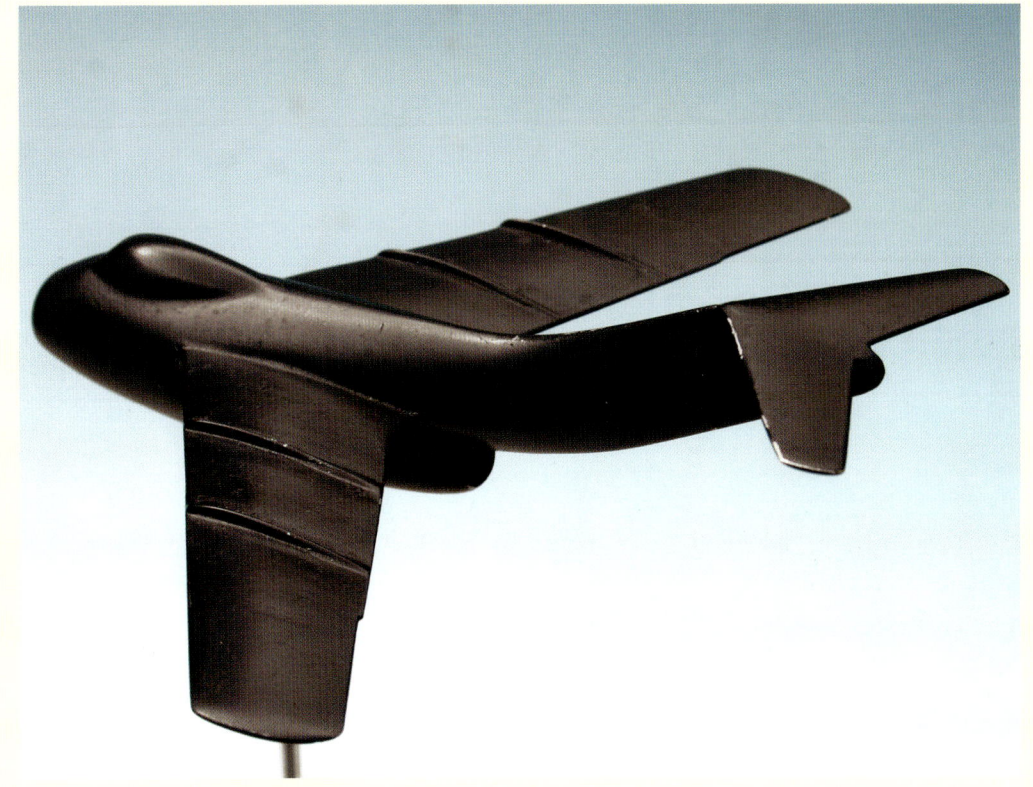

Here are three contemporary ID models: British, Dutch (NATO) and US, in wood, metal and plastic respectively — each representing Russia's first operational jet fighter, the MiG-9. All to the same scale, 1/72, they show differences in fuselage shape as well as significant differences in size. They also show the scarcity of reliable intelligence. *(George Cox)*

The Douglas XA2D Skyshark was the Navy's attempt to come up with a powerful contra-rotating turboprop attack plane. Unfortunately, engine-maker Allison could not get its propeller gearbox to work correctly. (Author)

Chapter 4
Fly Navy!

INTER-SERVICE RIVALRY has long been a routine feature of military life. Indeed, it is inevitable given the continuous need to fight for funds and influence. However, it became particularly focused in the post-war years when the newly formed USAF and the US Navy (which had shown the value of carrier-born airpower) both saw themselves as the way of projecting US military power around the world.

Political power struggles within the Pentagon centered on the case for developing manned long-range bombers against that for building bigger and better aircraft carriers.

Given these competitive pressures it was not surprising that the Navy took a keen interest in the prospect of jet aircraft. During World War II, it had carefully observed the construction of the Bell XP-59A and had obtained two YP-59As for its own evaluation purposes. In theory, jet-powered aircraft had a lot of advantages when it came to carrier operations – beyond the obvious potential to fly at higher-speeds and higher altitudes. Because the engine did not have to be placed in the nose, the pilot could be positioned to give far better forward visibility; nose-wheel undercarriages (which were more readily fitted to jet aircraft) were much better suited to deck landings; jets did not produce the massive torque when the throttle was opened (as happened with powerful piston-engines driving large propellers, causing the aircraft to roll in the opposite direction); and kerosene, being far less flammable, was a safer fuel than petrol.

A model of the Ryan XF2R-1 Dark Shark (left) stands alongside the Ryan FR-1 Fireball (right). A much better performing aircraft, the Dark Shark was a modified FR-1 with increased tail surfaces and a new nose that held a General Electric TG-100 turboprop engine driving an 11ft (3.4m) four-bladed Hamilton Standard prop. Both used the same General Electric I-16 turbojet in the tail. (San Diego Air & Space Museum)

Thus a jet-powered interceptor was viewed as very desirable by the Navy. But its early investigations were not all positive. The data gathered from the British Gloster E.28/39 and the XP-59A indicated that jet aircraft needed excessively long take-off runs, possessed high touchdown speeds, had slow throttle response time and had a voracious appetite for fuel – hardly a great set of characteristics for a sea-going airplane!

Ryan XFR-1 Fireball and XF2R-4 Dark Shark

The Navy therefore took a step back and started to look at the idea of dual-powered aircraft, planes that could utilize the advantages of both propellers and jet power. The prop engine would be the main power source and the jet would provide boost in climb and in combat. Admiral John S. McCain (father of Arizona Senator John McCain) called for proposals in December 1942 for a single-seat shipboard fighter combining a piston engine and a turbojet. Of the nine respondents, Ryan was selected with its Model 28 design and given a contract on February 1943 for three prototypes designated XFR-1.

The primary powerplant for the XFR-1 was a 1,350hp (1006.7kW) Wright R-1820-72W Cyclone nine-cylinder air-cooled radial engine. This was used for takeoffs and landings. And in the back of the XFR-1 was a 1,600lb (7.1kN) thrust General Electric I-16 (later re-designated J31) turbojet for boost in climb and combat. On 25 June 1944, the first prototype made its maiden flight, albeit without the jet engine installed. During the following month this was rectified and the aircraft it was flown with both engines in place.

However, by the time the first squadron of Ryan FR-1 Fireballs (its new designation) had finished qualifying for carrier operations, the Japanese had surrendered. After the war, the Navy canceled the order for the FR-1, but Ryan continued improving the Fireball. The FR-4 had modified air intakes now on the fuselage and replaced the turbojet with a Westinghouse J34 which increased top speed by 100 mph (160.9km/h). But the Navy was no longer interested.

Ryan continued developing the Fireball which ended with the XF2R-1 Dark Shark. It swapped the three-blade propeller and its piston engine for a General Electric TG-100 turboprop engine with an 11ft (3.35m), four-blade Hamilton Standard propeller and kept the General Electric I-16 turbojet in the tail. The new turboprop made the aircraft 4ft (1.22m) longer than the Fireball, and gave the aircraft a long, sleek bullet-like appearance. Although this was a modified FR-1, its increased tail surfaces and new nose gave the effect of a very different-looking machine. Its first flight took place on 1 November 1946.

The XF2R-1 Dark Shark offered much higher performance but the Navy was now firmly committed to pure jets for its future aircraft. However, the Dark Shark did catch the eye of the USAAF which had observed it at Muroc Flight Test Center (now Edwards AFB) in late 1946. At that time the Army Air Force had been evaluating its own composite fighter the Convair YP-81, though with disappointing results.

The YP-81 program was officially terminated on 9 May 1947 but the USAAF asked Ryan to modify the XF2R-1 by replacing the General Electric I-16 with the Westinghouse J34 turbojet engine instead. This new design was designated the XF2R-2 and proved to have good range and performance, though with only the same climb rate as the FR-1 at lower altitudes, 5,000ft/min (1,524m/min). The Dark Shark was an impressive composite aircraft but it would never rival the potential of a pure jet. Eventually the USAAF would put it aside as well.

Above: *The front powerplant for the FR-1 would be a 1,350hp Wright R-1820-72W Cyclone nine-cylinder air-cooled radial engine used for take-offs and landings. In the tail would be a 7.3kN General Electric I-16 (later re-designated J31) turbojet for boosting power in the climb and in combat.* (Allen Hess model)

Ryan tried to continue the relationship with the Navy after the Fireball was terminated. So Ryan modified the FR-1, including extension of the forward fuselage to accommodate the TG-100 turboprop and came out with this hopped-up version called the Dark Shark. (San Diego Air and Space Museum)

Fly Navy! | Chapter 4

Navy Seeks Advanced Design

The Navy's interest was in an aircraft with a wide range of speed; slow enough to safely land on a pitching deck and fast enough to match enemy aircraft in combat. It also saw a need for a high-speed interceptor that could be dispatched quickly and meet enemy bombers long before they had a chance to reach their target. These demands caused it to issue a set of requirements to the aircraft manufacturers that would result in some highly imaginative designs.

Vought V-173

The Vought V-173 was essentially an oval wooden airfoil with fabric covering, spanning 23ft 4in (7.1 m) with a length of 26ft 8in (8.1m). It had two large, three-blade wooden propellers 16ft 6in (5m) in diameter. The craft's short takeoff and landing ability was astounding. It could be airborne in only 200ft (60.9m) and if there was a 25mph (40.2km/h) headwind it would simply levitate in place and land with no ground roll. This aircraft could not be put into a spin or stalled and it had the ability of performing sustained flight at a 45° angle-of-attack while flying at 34mph (54.7km/h). The V-173 made its first flight on 23 November 1942. Not surprisingly it was referred to as the "Flying Pancake."

The war relegated the V-173 to low priority, but up to 1947 it flew 131 hours in 200 flights and provided good engineering data.

Right: The Vought V-173 "Flying Pancake" was a proof-of-concept aircraft to show that the design could fly; although the company had already made scale flying models. The landing gear was fixed and just like most modern fighters of today the pilot needed a ladder to get into the cockpit as depicted on this 1/72 Sword model. *(Barry Webb model)*

Below: This scratch-built 1/48-scale model of the Vought XF5U-1 "Flying Flapjack" represents a larger, more powerful version of the V-173 ordered by the Navy. At one point there were discussions of having the rotors articulate up, somewhat like an Osprey, to provide VTOL capabilities. In a way, it was the Navy's first inkling to what would eventually become the helicopter. *(Allen Hess model)*

Vought XF5U-1

During the testing phase of the V-173 the Navy called for a larger version, the XF5U-1. Two were built, one for static tests and the other to be the flying model. This new version had more powerful engines (ultimately driving four-blade props which rotated in opposite directions), metal skin, and retractable landing gear. The engines were two 1,350hp (1006.7kW) Pratt & Whitney R-2000-7 Twin Wasps and the craft was expected to fly as fast as 500mph (804.6km/h) and yet be able to slow down to just 40mph (64.4km/h). This XF5U-1 was now named the "Flying Flapjack" and its construction was completed on 25 June 1945. It underwent high-speed taxi runs but never actually flew. Sadly, just before its first flight, the program was canceled on 17 March 1947. Time had passed and the project was way behind in its development, and jets were now starting to come into service.

As the V-173 and the XF5U-1 were being developed and tested, WWII was coming to an end. The desperate Japanese Kamikaze suicide attacks sparked the need for an interceptor that could climb quickly and destroy the incoming enemy. Just before the war, the Navy had been looking into a Grumman two-engine fighter aptly named the Skyrocket. Although this aircraft dates from just outside the time period covered by this book, it is included to show the progression that eventually led to the F7F-1 Tigercat.

Left: Half of this scratch built metal 1/32-scale model of the Vought XF5U-1 "Flying Flapjack" is painted in its Navy colors, while the other half is left in natural aluminum. Compare this view with the overhead of the V-173 below to get an idea of how the body shape changed between the two versions. *(Author)*

Below: Top view of the Vought V-173 "Flying Pancake" would make anyone wonder if it could actually fly. The proposed V-173 was to have the pilot fly in a prone position, but the actual flying V-173 was changed to a standard seating position. Vacuform Model 1/72-scale *(Barry Webb model)*

Grumman XF5F-1 Skyrocket

The Grumman XF5F-1 Skyrocket was a very advanced looking design that promised all the benefits of a superior interceptor; lightweight, with two powerful engines enabling it to reach altitude quickly. The prototype first flew on 1 April 1941.

This aircraft proved to be quite capable, yet Grumman failed to gain any production orders. The Army Air Corps took an interest in it and asked for some modifications and a different version with the designation XP-50 was built for the USAAF. It had a longer nose to accommodate more weapons and tricycle landing gear, but it too was passed over.

Grumman continued researching and developing a better version of the XF5F-1 on its own until the end of 1944. Eventually, this design was resurrected as the Grumman G-51 proposal, which would become the F7F Tigercat.

An interesting model was found at the Grumman Archives representing a two-engine proposal that may well be an early form of the Tigercat. Its identification card stated that it was used in development studies based on the Skyrocket.

Top right: The execution of Grumman's in-house wood model of the XF5F-1 Skyrocket is top quality. Grumman and North American were considered the best model makers. Painted in pre-war colors using a high gloss lacquer, even the individual window panes have shading to give it a meticulous sense of detail. *(Northrop Grumman History Center)*

Right: The Grumman in-house wood model of the XP-50 was painted in an unusual shade of blue for an Army Air Forces proposal. This model shows the large canopy carved from a single piece of Plexiglas. It also carries the pre-war red-and-white striped rudders. *(Northrop Grumman History Center)*

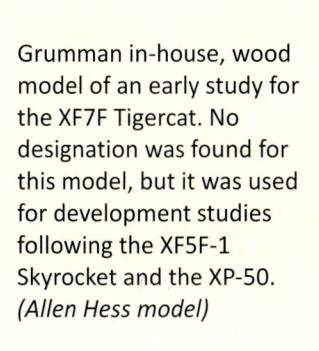

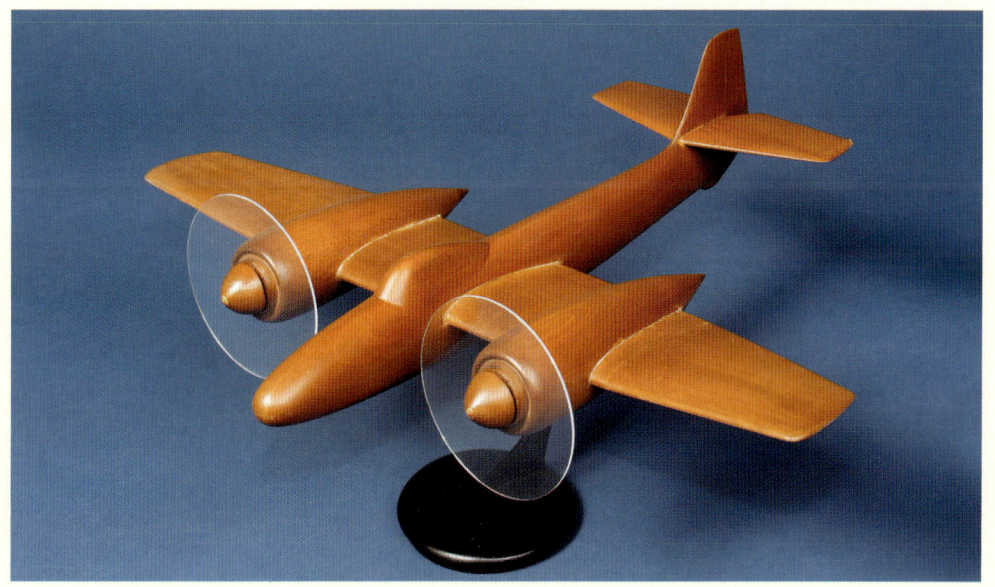

Grumman in-house, wood model of an early study for the XF7F Tigercat. No designation was found for this model, but it was used for development studies following the XF5F-1 Skyrocket and the XP-50. *(Allen Hess model)*

Chapter 4 — Secret US Proposals of the Cold War

In-house Grumman wood model of Design 75 dates from September 1945. An early study for a two-seat, four engine, carrier-based night fighter that was offered to the Navy. It appears to come from the Tigercat design. *(Northrop Grumman History Center)*

This unknown Grumman model is an early jet fighter study. It could possibly be one of the first design studies for the Grumman Panther. *(Northrop Grumman History Center)*

Early concept study possibly for the F11F Tiger. This in-house model found at the History Center had no designation but it does have similar looking nose, canopy, and inlet designs to the Tiger. *(Northrop Grumman History Center)*

Republic AP-46: The Turboprop Blues

During the time the Navy was stymied on jets for aircraft-carrier operations, an attractive alternative for them was the turboprop fighter. Such a craft could offer jet-type performances with advantages of a propeller-driven aircraft: power to take-off with heavy loads in short distances, sustain high speeds, and the ability to land at slower speeds or quickly power-up for go-arounds.

The Allison Division of General Motors began developing a series of turboprop engines for the Navy at the close of World War II. One of the more promising was the T40, with twin gas turbines connected to a reduction gearbox that could turn a propeller at supersonic speeds. The Navy had high hopes for the T40 and installed it in various projects including the North American XA2J Super Savage, the Douglas XA2D Skyshark, the Lockheed XFV-1 Salmon, Convair XFY-1 Pogo, the P5Y and R3Y Tradewind flying boats.

In 1951 the Navy requested a turboprop fighter and joined in on an Air Force program already investigating this aircraft type along with supersonic propellers. Republic proposed the AP-46, later designated XF-106, but eventually changed to the XF-84H. Three aircraft were ordered but only two prototypes were built. Modifying two F-84-35-REs to accommodate Allison's T40-A-1 turbine engine, Republic elongated the fuselage, and configured the empennage into a T-tail in order to ride above the propeller-induced turbulence. To avoid the complexity and trouble-prone gearbox associated with contra-rotating props, only a single three-blade stub propeller was used. The XF-84H's turbine engine would run the propeller at a constant high-speed, and the throttle control would change the pitch of the propeller blades to give it thrust. It could also reverse the blades, acting as a brake while taxiing.

But development problems persisted with the T40 delaying the first flight which finally took place at Edwards AFB on 22 July 1955. One unforeseen problem during ground run-ups was when the 12ft (3.7M) diameter stubby propellers reached supersonic speed, it gave off a hideous, high frequency squeal that induced nausea and headaches in the ground crew and pilot. The former Thunderstreak was re-named the "Thunderscreech."

It was reported that the aircraft handled well in flight, but not the Allison T40. Out of 12 test flights made in total, 11 resulted in emergency landings. However, the XF-84H did break one record that stands to this day, it did achieve 670mph (1078.3km/h): the highest speed attained in level flight by a propeller-driven aircraft. Despite that accomplishment, the XF-84H was canceled in late 1956.

Republic's wood model of the AP-46. Although the Navy requested this turboprop conversion, it was sponsored by the USAF Wright Air Development Center. Thus the USAF markings were retained on the model and the prototypes. (Cradle of Aviation Museum)

Chapter 4 — Secret US Proposals of the Cold War

Above: The XF-84H prototype never used the contra-rotating props. Note the black anti-torque fin behind the cockpit that countered the torque from the supersonic 12ft (3.7M) diameter propellers. Scratch-built 1/48-scale model. (Allen Hess model)

Close-up of Republic's proposed AP-46 model. It appears that the Republic model shop modified a YF-96A model by cutting off the nose, adding the large spinner with contra-rotating props, and the flush intakes. (Cradle of Aviation Museum)

Douglas XA2D-1 Skyshark

The XA2D program began in 1947, when the Navy's Bureau of Aeronautics released its final requirements for a new attack airplane with a combat radius of 600nm (1111.8km) and which could operate from escort-class carriers. Douglas received a letter of intent in June 1947, after it had been declared the winner of the competition, and orders were placed for two XA2D-1 prototypes by September.

First flight of the Skyshark was to be accomplished by March 1949. It was originally designated the AD-3, and to be powered by a Wright turboprop. After some major design changes the Skyshark was re-designated A2D which including the replacement Allison T40. With the new engine the horsepower jumped from 2,700 (2013.39kW) to 5,500 (4101.4kW). But the program suffered great delays because of the T40. By 1950, it had made 14 flights and, after one fatal crash caused by engine failure, the A2D did not fly again until 1952. Again, two more in-flight engine and gearbox failures in 1953 and 1954 ended the program. Douglas, frustrated with the inability of Allison to deliver a reliable engine, wrote to the Navy recommending that the program be terminated. In September 1954, Douglas's wish was granted and the Skyshark's career was over.

One of the by-products of the A2D program was the development of streamlined bombs that could be carried by high-speed aircraft. The old bomb shapes of World War II created too much drag, and the fins developed a habit of buffeting and either breaking or bending,

Because the Douglas A2D Skyshark was to be a carrier-based aircraft, folding wings were required as seen in this large in-house wood model prepared by the Douglas model shop. (Author)

The formidable front end of the Douglas XA2D Skyshark features its contra-rotating propellers. Also under wing are the streamlined external stores that were developed by Douglas to help reduce drag. (Allen Hess model)

adversely affecting the bomb's trajectory. Douglas was contracted to develop stabilizing fins, new bomb shapes, and arming devices.

These studies were also applied to other external stores, and flight tests using re-designed bombs and fuel tanks on the XF3D Skyknight revealed a substantial improvement in performance. Extrapolating this information to the A2D, it was estimated that the Skyshark would fly over 50 knots (92.7km/h) faster when carrying three redesigned bombs

This view of the Douglas XA2D Skyshark shows its resemblance to the AD-1 Skyraider, yet they were very different. Had the Skyshark's contra-rotating turbine engine worked, it would have been a formidable attack aircraft.. (Allen Hess model)

Attack Aircraft, Fighters and Flying Boats

As jets became more advanced, new techniques were developed to facilitate their operation at sea. Two of the most significant were ideas adopted from the British Royal Navy: the angled deck and the mirror-based approach aid. These allowed higher performance aircraft, with much higher take-off and touch-down speeds to be flown on and off carriers. The jet thus became the predominant fixture in the Navy's arsenal. Nevertheless, the Navy still found a use for reciprocating engines and turboprop power-plants, which continues right up to the present day. And between 1945 and 1965 it looked at many unique proposals, including flying boats and nuclear propulsion.

Attack Aircraft – Big Plans for a Big Carrier

The following attack aircraft designs were submitted to a design competition won by the Douglas A3D Skywarrior – in response to Specification OS-111 calling for a Class VA Heavy Attack Aircraft that could be developed and built using current available technology. There was also OS-115 which called for a much more advanced, high-speed aircraft.

In mid 1945, the Navy started to investigate the potential use of large aircraft carriers, super-carriers actually, with specially designed attack bombers. There is a good book on this subject matter written by Jared A. Zichek ('The Incredible Attack Aircraft of the USS *United States*, 1948 – 1949').

It was argued that these carriers resolved the problems that the Air Force had with flying long-range land-based bombers that could reach the Soviet Union. This was also a political move to keep the Navy engaged strategically as part of the nuclear deterrent.

By August 1948, when it seemed that the construction of these new flush-deck carriers was going to be approved, the Navy BuAer issued Outline Specifications to the aircraft manufacturers, OS-111 and OS-115. OS-111 called for a high-performance attack aircraft that could be made with the current technology available. OS-115 requested a far more advanced, high-speed, long-range attack aircraft.

On 29 July 1948 the Truman administration approved funds for the construction of five "super carriers." The *USS United States* was to be the first of the largest carriers ever built. It was termed a flush-deck carrier, for the ship would have no command tower structure or island above the flight deck. This new deck would be designed for the exclusive use of aircraft operations, specifically for a new breed of long-range nuclear bombers. The *USS United States* could conduct different flight operations simultaneously: landings would be carried out in the rear deck while launching aircraft from mid and bow catapults. The Navy would now become a big part of the nuclear offensive capability of this country.

But this new, larger nuclear role for the Navy threatened the Air Force's monopoly on strategic nuclear weapons delivery, and with it, its large share of government financing. The issue quickly became a political hot potato and the Air Force managed to stop the program just five days into the keel construction of the USS *United States*.

While the program was still alive, the aircraft manufacturers designed and proposed new ideas to the Navy specifically for this aircraft carrier, to both OS-111 and OS-115.

Fairchild Model M-128

Fairchild's M-128 design was made in response to the BuAer Outline Specification OS-111 issued to aircraft manufacturers which reqested a heavy attack bomber that could carry a 10,000lb (4.536kg) nuclear bomb internally. The aircraft would be over 95ft (28.9m) in span, including the tip-tanks, and over 60ft (18.3m) long and it was powered by 6 Westinghouse XJ46-WE-2 turbojet engines. The fuel issue for this aircraft was uniquely addressed by adding a second wing. The M-128 was designed to fly two wings; in some respects like World War I aircraft. The top wing was a flying gas tank, carrying the extra fuel it needed, and once the fuel was spent the wing was simply ejected in flight, allowing the aircraft to carry on with the mission with enough gas left over to get it back to the carrier.

An illustration of the Fairchild M-128 showing the attack bomber releasing its upper wing that carried extra fuel. On the ocean below can be seen the aircraft carrier that launched this aircraft. (National Archives via Ryan Crierie)

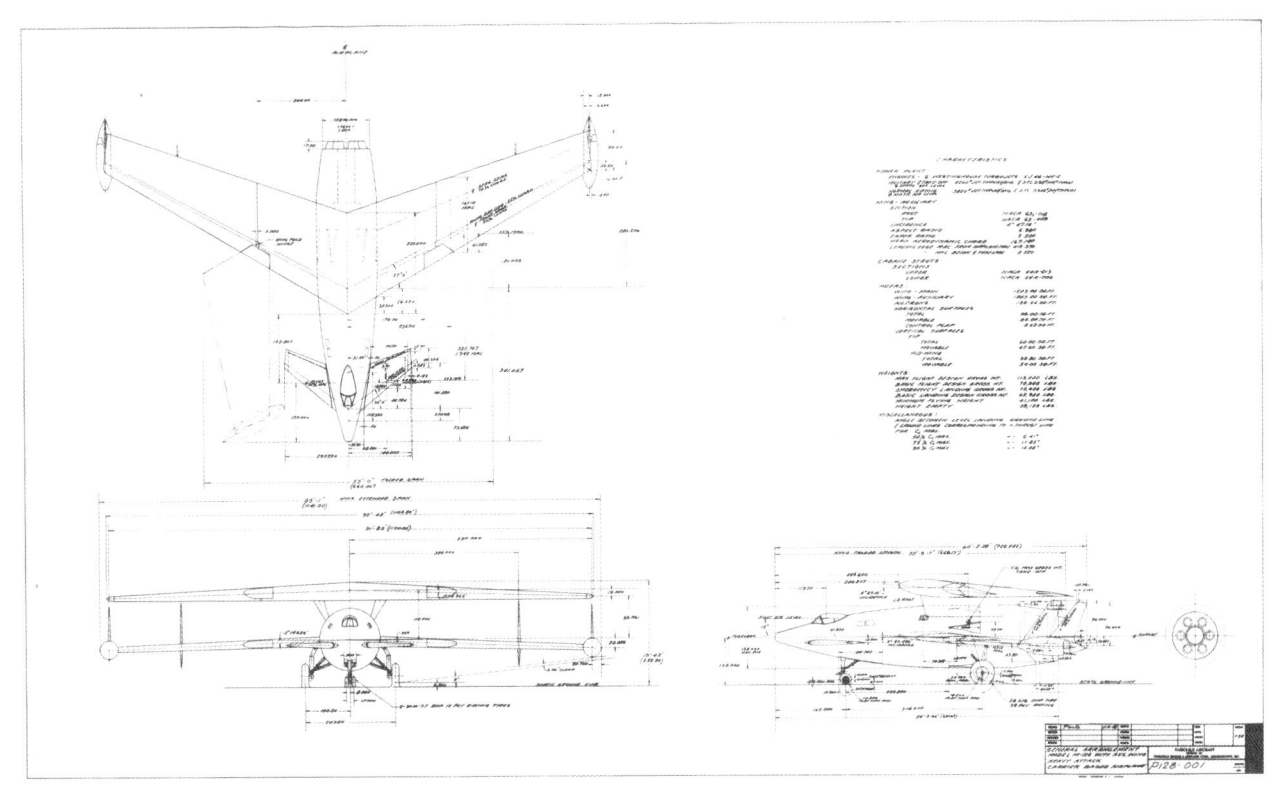

Three-view of the Fairchild model M-128. This heavy attack bomber would deliver a nuclear bomb by flying over the target and simply ejecting it out of a rear portal. Seen in the drawings is the upper wing/fuel tank that could be jettisoned allowing M-128 to fly back to the carrier with its single lower wing. (National Archives via Ryan Crierie)

Of note was the delivery system of the bomb itself. Once over the target, it simply ejected the bomb from its rear fuselage. This same bombing technique was later employed in the North American A3J Vigilante attack aircraft.

Lockheed L-187-7

Although this is not a pure jet, the L-187-7 is included for it was an OS-111 proposal from Lockheed. This attack bomber had a 35° swept-back wings and was powered by two Allison turboprop engines swinging contra-rotating props.

The Standard Aircraft Characteristics states the following in the MISSION AND DESCRIPTION section:

"Execution of high-altitude horizontal bombing attack mission with large bombs and special weapons of 10,000lb (4,536kg) weight. Designed with the specific mission of high transonic speed long-range 'hit and run' attack and to be operated off and on the new class CVB flush-deck carriers."

"Configuration features sweptback wing with two triple-turbo-prop engines geared into 12 blade dual rotating propellers, crew of three in pressurized cabin with access to bomb bay in flight, and fully retractable dual-tire bicycle landing gear with outriggers mounted on and retracting into the nacelles."

"It has a tail turret, Aero XII E type with APG-25 radar, and two M-25 20mm guns with 500 rounds each."

Wingspan	91ft 6in (27.9m)
Length	83ft 7in (25.3m)
Height	24ft 10.5in (7.3m)
Power plant	Two Allison XT-44-As

The 35-degree swept back wings with the contra-rotating Alliso XT44-A turboprop engines gives this drawing of the Lockheed L-187-7 from the Standard Aircraft Characteristics folder a powerful look. (National Archives)

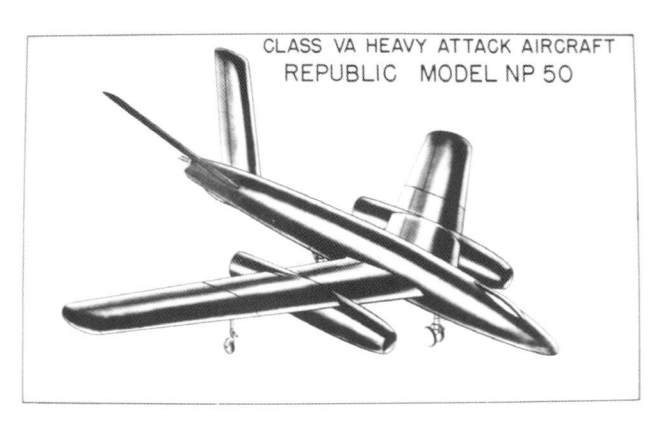

An illustration of the NP-50 from the Standard Aircraft Characteristics folder provides a glimpse of Republic's signature V-tail and swept-back, slightly tapered wings that make this attack bomber look sleek and fast. (National Archives)

Republic NP-50

The NP-50 attack aircraft proposal was a beautiful, sleek design which included Republic's signature V-tail. Had it been built it would have made for an impressive aircraft. It was offered to the Navy in November 1948 in response to OS-111. The Standard Aircraft Characteristics folder found at the National Archives in Washington D.C. reports the following in the MISSION AND DESCRIPTION section:

"The Republic NP-50 aircraft is a four engine jet-powered aircraft employing a swept-back variable incidence wing, V-tail, and dual-wheel tandem type gear. The engines are housed in two nacelles supported under the wing structure. The bomb bay is designed to carry a special bomb load of 10,000lb (4,536kg). The primary mission of this airplane is the execution of a high altitude, horizontal bombing attack with large bombs and special weapons at a point not less than 1,700nm (3150.1km) from the point of take off."

The tail has a radar-controlled AGL turret with two 20mm M-24 guns, 500 rounds each, and these specifications:

The difference between this in-house 1/50-scale wood model of the Republic NP-50 and the illustration above is the engine configuration. On the model, twin engines are stacked vertically in each nacelle. Also, the canopy is much longer on this model. (Cradle of Aviation Museum)

Length	96ft 2in (29.3m)
Span	85ft (25.9m)
Height	25ft (7.6m)
Power plant	Two Westinghouse XJ40-WE-12s
	Two Westinghouse XJ46-WE-2s

Douglas El Segundo and Santa Monica OS-115 Designs

Although this book has excluded the research-dedicated "X" planes, the Douglas Model 499C/Project MX-656, better known as the X-3 Stiletto, deserves a mention because of its influence on various military designs.

It was designed for one purpose, high-speed research, and it first flew on 15 October 1952. It had short, stubby wings on a long, slim fuselage, and was powered by two Westinghouse J34-WE-17 engines. As with so many airplanes of the time, the available power-plants were simply not capable of giving the aircraft the capability to fulfill its design performance.

On the positive side, the aircraft did provide information on the relationship between aerodynamics and load distributions. It also pioneered the use of titanium components. Douglas proposed X-3 variants to the Air Force for its parasite bomber (see Chapter 2), and to the Navy as the Model 499C, a long-range high-speed attack aircraft for the up-coming *USS United States* aircraft carrier.

In fact two Douglas design teams – at El Segundo and at Santa Monica – came up with responses to OS115: Model 594 and the Model 1186, both of which used aspects of the X-3's design.

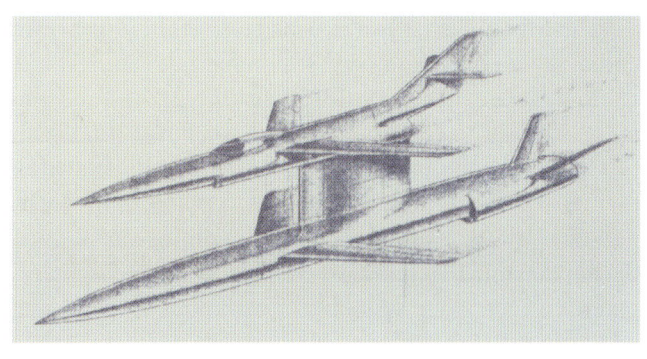

Illustration from the Standard Aircraft Characteristics folder of the Douglas Model 594A from October 1948. This shows a variation of the jet with a large missile. (National Archives)

Illustration of Model 1186-1 from the Standard Aircraft Characteristics Folder, October 1948. This was an advanced version of this attack concept. The aircraft shown here looks very much like the Stiletto. This has a much larger weapon with four jet engines powering the missile. (National Archives)

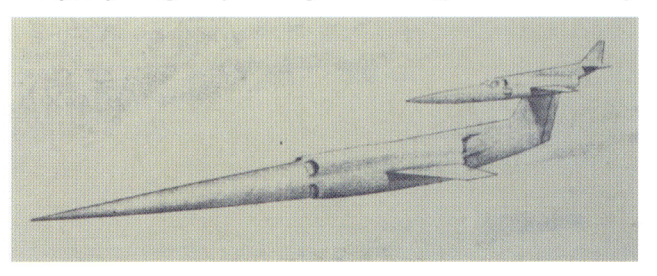

They were a modern twist of the Mistel concept the Germans tried during World War II which consisted of a Focke-Wulf 190 A-8 or the Bf 109 F-4 attached to the top of a Junkers Ju 88 A-4 bomber, laden with explosives. The pilot of the fighter guided the bomber to its target, released it and flew home; except in this case the X-3 variant carried a powered missile.

A revolutionary design when it first flew in 1952, the Douglas X-3 Stiletto provided valuable research data on stub wings and the inertia-coupling phenomenon. The X-3 was also the first airplane to use titanium structure and have an air-conditioned cockpit. (Allen Hess model)

The Model 594 was a two-place aircraft with a pilot and a navigator/radio operator/bombardier. The aircraft itself weighed just over 18,500lb (8392.5kg) and the entire composite craft, with bomb, came in at 100,000lb (45,360kg). This external, expendable vehicle which contained the bomb and fuel was a plane unto itself, powered by twin turbojets. All the control surfaces of the expendable container were connected to the fighter's controls. There were a number of variations of the Model 594 with different wing and tail surface designs.

The Model 1186 again incorporated the escape aircraft as a variation of the company's X-3 Stiletto. And like the Model 594, the Model 1186 was drawn into four various configurations of the expendable vehicle. It is interesting to note that this version has the escape aircraft mounted on the tail of the larger, bomb-carrying vehicle, thus utilizing the 1186 as the elevator for the larger craft.

There were a few differences between the Model 1186 and the X-3. The 1186 was to have one Westinghouse X24C-10, while the X-3 had two of the more powerful Westinghouse J34-WE-17 turbojet engines. The 1186 carried two crew-members, pilot, and navigator/radio operator/bombardier, while the X-3 had only a test pilot. The 1186 had a tail hook, and presumably a much strengthened undercarriage, for the return landing on the carrier. The X-3 was land based.

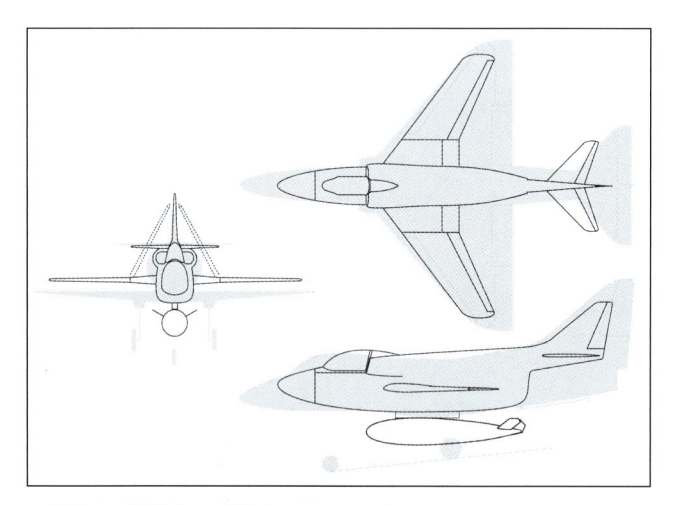

To accommodate the single engine, the 1186 had a smaller airframe in length and wingspan, shorter than the X-3. The 1186 measured 54ft 5in (16.6m) in length with a wingspan of 19ft 3in (5.9m). The X-3 was 66ft 9in (20.3m) long and had a 22ft 8in (6.9m) wingspan. The expendable vehicle was powered by four Westinghouse XJ40-WE-10 engines with afterburners.

Douglas Model D-640

There was another unusual attack aircraft model from Douglas, the small Model D-640 which in this case was to be carried by a submarine. There is not much information on this model, found at the San Diego Air and Space Museum. It was tucked away in one of the many boxes that house the Ed Heinemann Collection.

Ed Heinemann was the Chief Engineer for Douglas, and one of the top aircraft designers of his era. He had been interested in producing one of the lightest jet fighters possible, and was approached by the ONR (Office of Naval Research) in the late 1940s to provide a sub-based aircraft to "attack enemy submarines on the high seas." Eventually, in 1954, Douglas came out with the very light and very fast XA4D-1 Skyhawk carrier-based attack aircraft, and the D-640 may have been the precursor, for in certain ways it looks similar to an A4D, although slightly smaller in size.

In 1952 Heinemann made his unsolicited proposal to the Navy for this submarine-based, light, jet-powered attack bomber. The idea may have been the same as the Japanese Aichi M6A1 Seiran was to the wartime I-400 class submarines. Possibly the D-640 may have been intended to be a manned backup of the sub-launched Regulus missile, carried in a deck-mounted hangar and launched from a rail, assisted by JATO. The original D-640 was to be powered by a single Westinghouse J34-W-36 engine, and although the Navy accepted the proposal, no prototype was ever produced that resembled this model.

Top left: *Three-view shows size and shape comparison of the D-640 with the Douglas A4D-1 Skyhawk. Although the D-460 never flew, the Skyhawk became the longest-produced U.S. tactical fighter of its time.* [Tommy Thomason]

Left: *A model of the D-640 with wings folded. The plaque on the base reads, "Special Weapon Bomber, Submarine Based, NONR- 722(00)."* (San Diego Air & Space Museum)

Early four-engine Grumman proposal for a carrier-based jet fighter, Design 75 (G-75) designated XF9F-1. When cancelled by the Navy, Grumman offered Design 79, later designated XF9F-2. (Northrop Grumman History Center)

The Grumman in-house model of Design 79-D starts to exhibit the appearance of the Panther, and is listed as the XF9F-2 from 1946. (Northrop Grumman History Center)

Fighters

Grumman Early Jets and Supersonic Series

Grumman had a long relationship with the Navy that continued into the jet age. When the Navy issued its Request For Proposals in 1945 for a carrier-based, jet-powered, radar-equipped, two-seat night fighter, Grumman responded with Design 75 (G-75) capable of 500mph (804.7km/hr), at an altitude of 40,000ft (12,192m) able to detect enemy aircraft as far as 125 miles (201.2km) away. The contract went to Douglas for their XF3D-1 but the Navy ordered two G-75 prototypes, designated XF9F-1 as back-ups to the Douglas proposals.

The G-75 was to be powered by four Westinghouse 24C turbojets mounted in pairs, in mid-wing pods. It was to have a nose radome, and armament consisting of four 20-mm cannon. Eventually, development studies showed the G-75 inadequate and the contract was to be cancelled. But Grumman offered a single seat fighter they had been working on, the G-79 and persuaded the Navy to use the XF9F-1 contract for the G-79. On October 9, 1946, the Navy amended the contract, ordered three single-seat prototypes, and the first Panther flew 24 November 1947 using a Rolls Royce Nene turbojet. The Panther became the first Navy jet to enter combat when it went to Korea in 1950.

In 19 August 1952 the Bureau of Aeronautics sent out a Request for Proposals. The Navy called for a new carrier-based day fighter issued under Specifications OS-130. This brings the Navy into the supersonic era. Grumman responded with its Design 97 (G-97) in a competition that was eventually won by the Vought F8U Crusader.

Although the G-97 had failed to win the Crusader competition, it was not without merit and went on to evolve into the G-98 series which was eventually designated F11F and named Tiger. This aircraft also saw service with the Navy, though was more troubled and somewhat less successful than its predecessors. The following G-98D, and G-98S were all proposed developments of the F11F.

One last concept to include in this series is the Grumman Design 118 from the early 1950s, later designated the XF12F-1. It was proposed as a task-force defense fighter, an all-weather, missile-armed, carrier-based interceptor. By 1956 the XF12F-1 was cancelled in favor of the F4H Phantom II. Allegedly it was to be named "Lion" had it been bought by the Navy.

Unknown Grumman Naval Model

Among the collection of models at the Grumman Archives are a number of Navy jet models that have no identification. An interesting one is a blank wood model that has an asymmetrical canopy, similar to the British Navy carrier plane the de Havilland DH.110 Sea Vixen. The unusual horizontal stabilizer comes to a point that sticks out well forward of the vertical. It is somewhat reminiscent of the tail on the Grumman XF10F Jaguar swing-wing fighter prototype. If the Jaguar can be used to help date this model, it would come from between 1952 to 1954.

A Grumman in-house wood model conceptualizes a Navy fighter with an asymmetric canopy. Again, nothing is known about this model. It may have been a simple design study. (Northrop Grumman History Center)

Chapter 4 — Secret US Proposals of the Cold War

The Grumman G-118 was to be a carrier-based interceptor. Two prototypes were ordered in 1956 but before construction started they were cancelled in favor of the F4H Phantom II. (Northrop Grumman History Center)

Although this Grumman Design 97 appears to be an early F11F Tiger it was actually a proposal for the next supersonic day fighter to replace the Cutlass. That competition was won by the Chance-Vought Crusader. (Northrop Grumman History Center)

The Grumman G-98 was the design that would lead to the F11F Tiger. This in-house wood concept model was proposed by Grumman to the Navy in 1952 after the XF10F Jaguar was cancelled. (Northrop Grumman History Center)

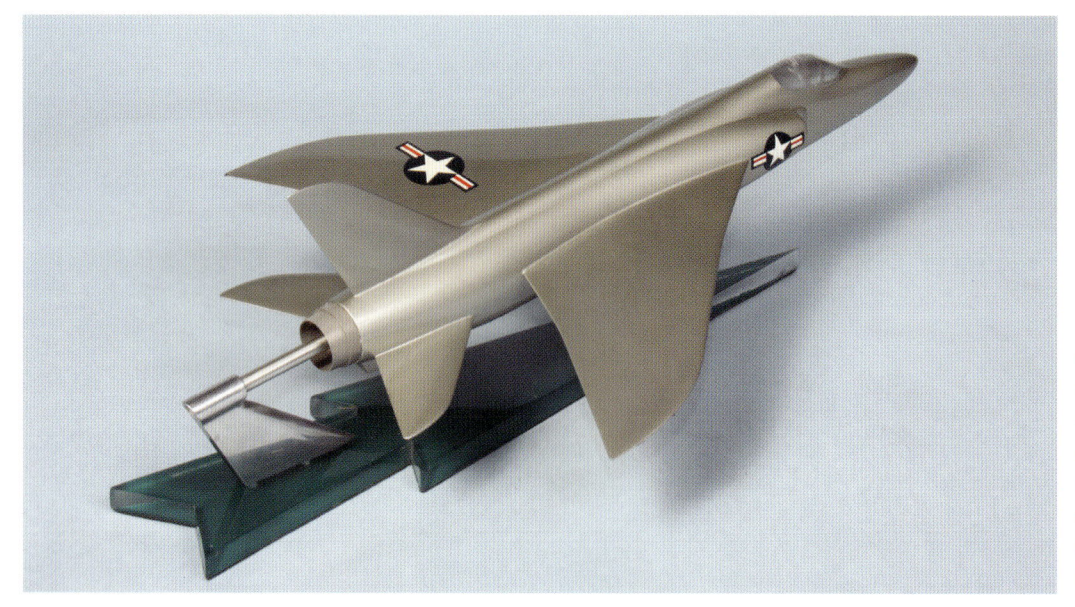

The Grumman F11F Tiger concept would go through a number of iterations. Here is the G-98D which goes back to slightly larger wings with more sweep. (Northrop Grumman History Center)

In this view of Republic's in-house wood model of the NP-48, one can see on the bottom of the forward fuselage the flush inlet to the turbojet engine. (Cradle of Aviation Museum)

The Republic NP-48 in-house wood model conceptualized the Navy version of the XP-91 with inverse-tapper wings and butterfly tail. The model has a rocket mounted on the lower rear fuselage. (Cradle of Aviation Museum)

Republic NP-48, NP-49

The NP-48 and NP-49 ('NP' stood for Navy Project) were most likely Republic proposals for US Navy Specification OS-113, which was won by the McDonnell design of the F3H Demon. The NP-48 closely resembles the AP-31 (XP-91) covered in Chapter 3. It has the butterfly tail, inverse tapered wings, and just a single rocket engine in the rear fuselage just below the tail exhaust. This design looks sleeker. The fuselage appears longer with the unobstructed radome nose, the canopy, which has been moved forward and the extended tail pipe to house the afterburner. But the biggest difference between the AP-31 and the NP-48 is the air intake. The NP-48 has a "NACA scoop" flush air intake located under the forward fuselage.

There were two versions of the NP-49. At first glance, they appear to look very much the same. Both use the same V-tail and inverse wing arrangements and both have their engines mounted under-wing on pylons. However, the differences between the two iterations are quite subtle but can be seen in the different ways the glass is designed around the nose, and how the inlets work on the engine nacelles.

From a slightly different angle, of the same NP-49 model shows its Republic style V-tail. (Cradle of Aviation Museum)

Republic took a slightly different approach with this 1/20-scale in-house wood model of the NP-49. With turbojets on pylons and a bullet-shaped fuselage, the NP-49 hinted at the Bell X-1. (Cradle of Aviation Museum).

A second variation of the Republic NP-49 shows that the glass canopy is now extended to the nose and below. It is hard to tell from this head-on view, but the twin engine nacelles extend longer than on the previous model. (Cradle of Aviation Museum)

Flying Boats

Before the war flying boats had been developed on both sides of the Atlantic as the basis for long-haul passenger travel, their lack of need for prepared runways around the world gave them an obvious advantage. Once the war started, flying boats were utilized to help patrol the seas, to look for ships above or below the surface, and for search and rescue of downed pilots. The experience of fighting island-to-island against the Japanese in the Pacific, emphasized the need to be able to operate from forward areas without airfields. The flying boat therefore assumed a more aggressive role, and so after the war, interest inevitably turned to jet-powered attack and fighter seaplanes.

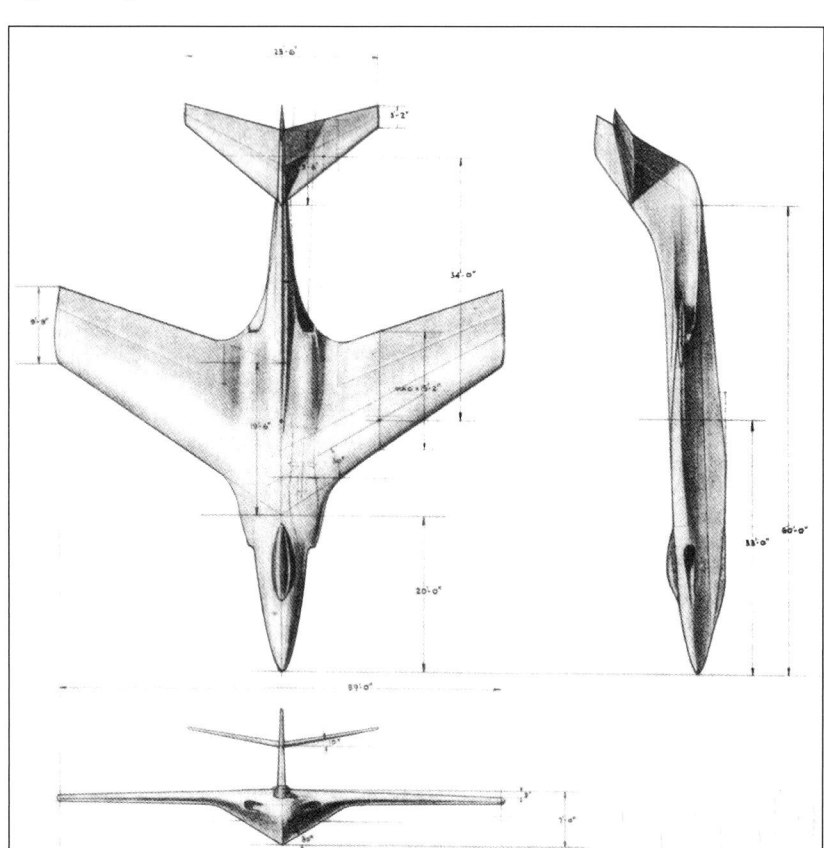

The overhead shot of the second variation of the Republic NP-49 clearly shows the nacelles as longer and rounded forms. The inlets for the engine nacelles are flush styles and can be seen on the front sides of nacelles. This in-house wood model is also in 1/20-scale. (Cradle of Aviation Museum)

Actually the British had already seen this coming and developed three prototypes in the late 1940s of a seaplane jet fighter, the Saunders-Roe SR.A/1. It performed well, although it was never ordered into production. In the US the military called upon Convair and Martin to study the possibilities. Martin came up with a large bomber type named the XP6M SeaMaster while Convair worked on the XF2Y Sea Dart hydro-ski fighter in the 1950s. The Sea Dart owes its existence to the hydrodynamic studies done by Convair in the late 1940s on a subsonic seaplane fighter model known as the "Skate."

The Skate came from a series of studies of swept-wing, shallow-blended hull, jet-powered seaplane designs. It never progressed past the model stage. Gradually, Project Skate evolved into Project Betta, which changed the design into a high-performance delta-wing seaplane. This wing form was also being explored by Convair on its XF-92A, and the XF-102 discussed in Chapter Three. It was therefore not surprising that the end result was a delta wing fighter, the Convair Sea Dart.

Artist's rendering of the Convair Skate proposal. This design was used to make a scale model for hydrodynamic studies in the late 1940s for a subsonic seaplane fighter. (Scott Lowther)

The Convair Sea Dart was based on the delta-wing XF 92A Dart. The studies from the Skate were used to help make this jet powered seaplane fly off the water. (San Diego Air and Space Museum)

Unknown Convair Flying Boat

While doing research at the San Diego Air and Space Museum, one of the Collections Curators, Al Valdes, was asked about a rumor that some unknown seaplane model was buried in the bowels of the museum building. He had never heard of such a thing, but said he would look. After two days of digging around, Mr. Valdes excitedly interrupted one of the model shoots with the news that something unique had been found. With a good washing of all the dirt and grime that covered this model, an unknown orange seaplane emerged. It seems to be a wind tunnel model of some jet version of the Convair P6Y patrol flying boat project with folded wings. The P6Y was proposed in the late 1950s as a three engine turboprop seaplane for anti-submarine warfare (ASW). This newly discovered model bears some resemblance to the P6Y but shows three turbojets. Other than that, no one knows exactly what it is.

1956 engineering drawing of Convair's P6Y ASW patrol seaplane proposal that was to be powered by three Wright R-3350-32W reciprocating engines. The drawing, shown for comparison, appears to have a close resemblance to the unknown turbojet seaplane model discovered at the San Diego Air and Space Museum. (Scott Lowther)

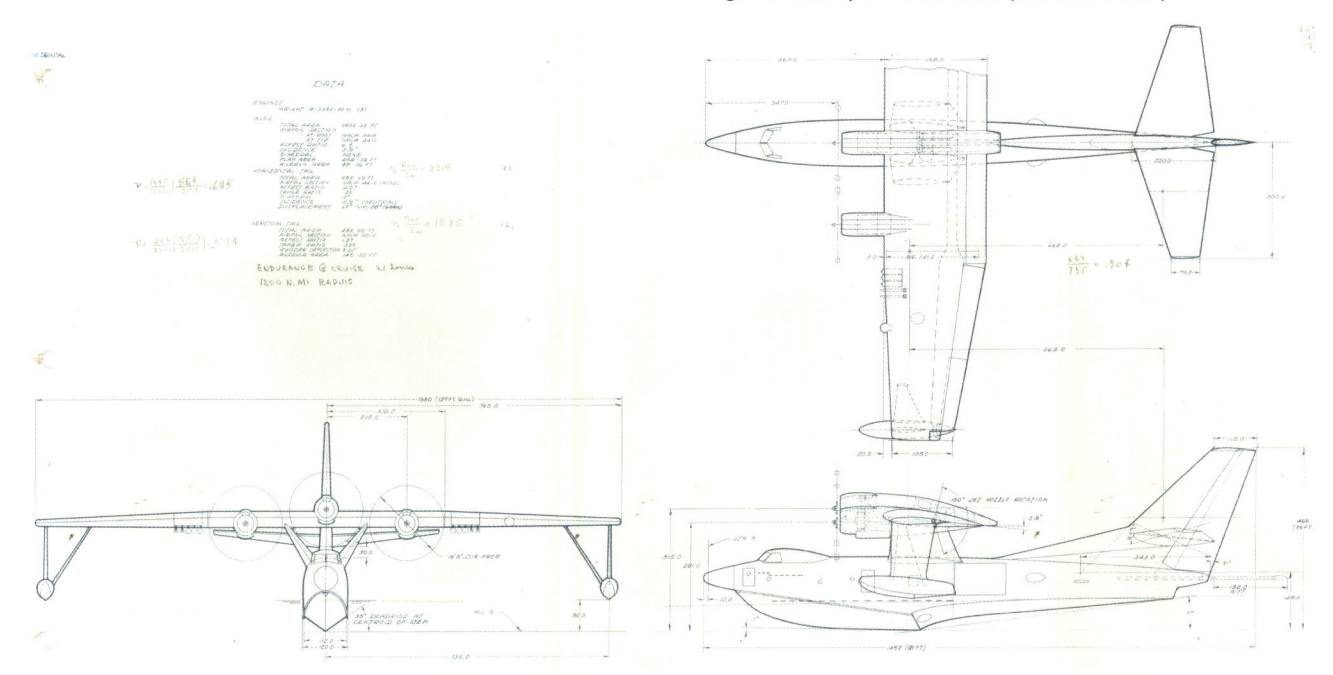

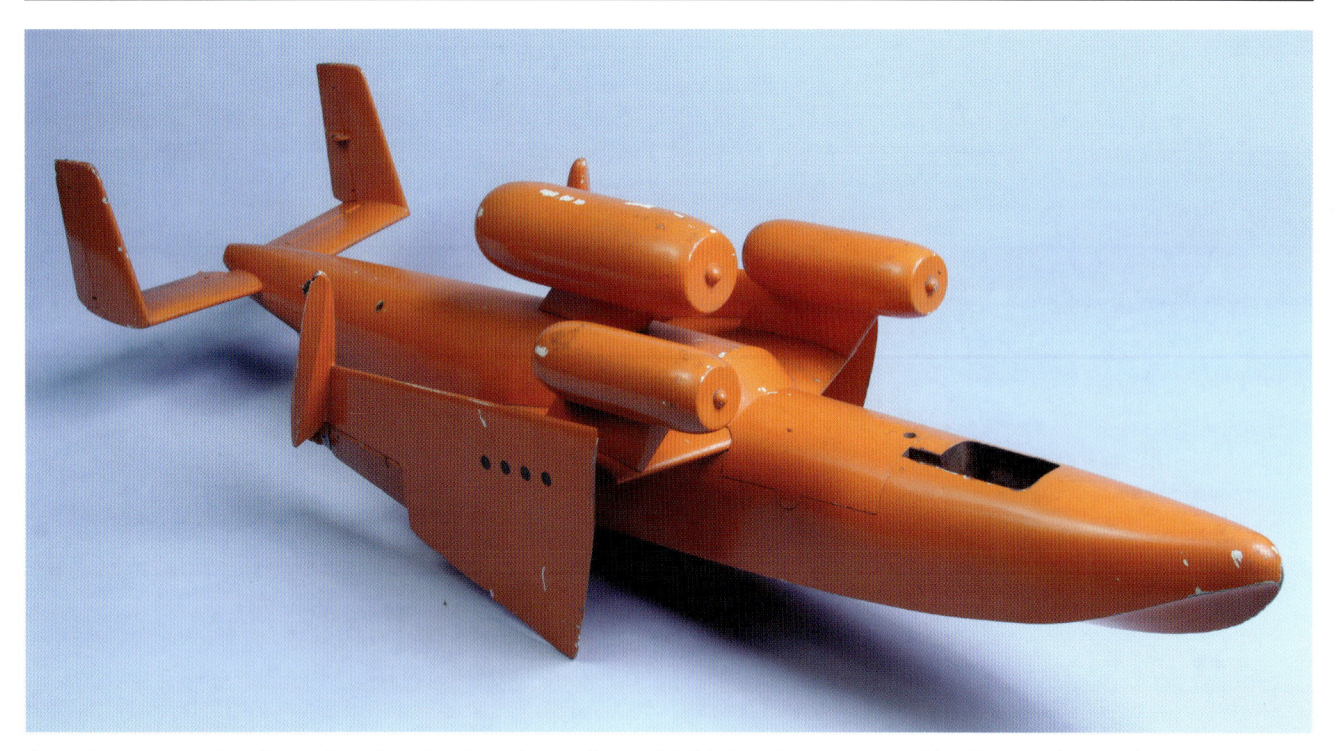

An unknown wood and metal model of a Convair seaplane with folded wings, powered by three turbojets. It appears to be a wind tunnel model but no information about it was available. (San Diego Air & Space Museum)

Nuclear-Powered Flying Boats

When the cry went out for nuclear-powered aircraft, the Navy made a very strong case for nuclear-powered seaplanes. In a 1957 article in the U.S. Naval Institute's *Proceedings,* Commander Arthur D. Struble, USN, Project Officer on Nuclear Seaplanes, Bureau of Aeronautics, states: "To help preclude (a nuclear) national disaster, the Navy is developing effective weapons for the defense of this country, one such weapon is the nuclear-powered seaplane." He quotes the then Secretary of the Navy, Charles Thomas: "If you want to romance its possibilities, look at a world map and imagine the numberless bases where a seaplane can operate. The oceans and the seas will be its bases. These water fields will cost nothing. They will require little maintenance and their use will pose no problem of sovereignty."

Getting to the point, Commander Struble declares: "With the nuclear powered aircraft, not only will the vast ocean areas become runways, but the striking force can be concentrated or dispersed quickly. With the range made possible for seaplanes by the use of nuclear fuel, the mobile bases supporting them can be moved farther back and spaced farther apart, thus decreasing the vulnerability of the bases to enemy attack. Furthermore, the nuclear-powered aircraft can find water and sea conditions suitable for its operations since an extra 1000 miles (1,610km) of cruising would make relatively little difference. The Navy's patrol, anti-submarine warfare, attack, and airborne early warning missions will require aircraft with long endurance and/or range capabilities. Nuclear-powered aircraft will satisfy these requirements. The *Nautilus*, the first atomic-powered submarine, has already demonstrated the advantages of nuclear propulsion by cruising thousands of miles without refueling."

Convair and Martin had already been working since May 1953 on Navy contracts to study nuclear-powered propulsion. Convair proposed to take an existing British seaplane, the Saunders-Roe Princess Flying Boat, and convert it into either a four- or six-engine nuclear-powered seaplane (as discussed in Chapter Two). By the mid 1950s Convair was working on nuclear aircraft propulsion on 3 fronts: it was already flying the land-based NB-36H while design of the NX-2 was about to start, and now it was involved in nuclear-powered flying boats. Convair made a number of design studies that looked like different-size versions of the Martin SeaMaster and its own Sea Dart, but with different wing configurations.

By the mid-1950s Convair came out with a series of nuclear powered sea planes studies, designated Model 23; there were four in this series, Model 23A, 23B, 23C, and 23D. Model 23A was configured with a T-tail and a delta wing having a span of just over 76ft (23.3m). The rest had swept wings of various designs. By 1958 wind tunnel tests were done on Model 23B, a larger proposed aircraft with 50-degree swept back wings spanning 115ft (35.1m) and

Convair's Model 23A was one of various model studies for a nuclear powered seaplane. This model has an opened bomb bay and the upper rear fuselage section comes apart to reveal the nuclear power plant. The model is made of wood. (John Aldaz collection)

a T-tail. It was designed to be a long-rang attack aircraft but could also do aerial mine-laying and photo-reconnaissance. The actual model of the 23B has removable engine covers, which show what appears to be a pair of Pratt & Whitney nuclear Indirect Air Cycle engines, similar to the engines used in Convair's NX-2.

An interesting aspect of these engines was the use of liquid sodium as secondary coolant for the reactor, which is highly volatile when it comes in contact with water. In fact, it explodes! Maybe this was not such a great idea for seaplanes after all.

By 1961 the concept of nuclear aircraft propulsion had come to a dead end. Too heavy, too slow, too much toxic pollution and too much expense with little results. The government stopped the funding and all large-scale investigations of nuclear powered aircraft ceased. Except the Navy would continue to use and develop nuclear power for their submarines and aircraft carriers.

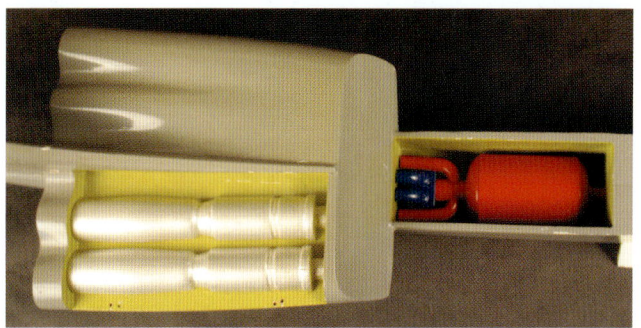

A close-up showing the installation of the nuclear reactor within the mid-section of the Model 23A. (John Aldaz collection)

The Navy had done well for themselves in the initial exploration and development of jet aircraft and the emerging forms of turbojet and turboprop propulsions. They now had powerful jet fighters and attack bombers that were all carrier-based. Throughout the 1960s the Navy would continued to work with manufacturers to develop even more advanced aircraft. Also beneficial for the Navy would be the new developments in vertical designs – helicopters, and in the evolution of drones.

From this angle the nuclear powered Model 23 by Convair displays it sleek lines. The swept-wing was to have a wing span of 76ft (23.3m). (John Aldaz collection)

Chapter 5
Vertical Flight and Other Concepts

THE PREVIOUS FOUR CHAPTERS examined aircraft that were meant to fly "normally." Needing forward speed to generate lift, they had to use runways (or long stretches of water) to accelerate for take-off and to decelerate on landing. This chapter discusses a range of designs, some never covered before, that attempted to do away with such constraints and simply take to the air vertically. The strategic possibilities of having a fighter aircraft that did not need a runway held great interest for the military. In battle, this type of aircraft would have certain advantages over conventional aircraft. They could operate closer to the front line, would need minimal field preparation, and give quicker response times with shorter turn-around times for re-fuelling and re-arming.

So with these new concepts came new terminology into the military vocabulary, VTOL and V/STOL or STOVL.

VTOL: vertical take-off and landing, describes the ability of an aircraft to operate without any runways whatsoever. What immediately comes to mind with VTOL aircraft are such turboprop-driven aircraft as the Lockheed XFV-1 Salmon, or the Convair XFY-1 Pogo, or the turbojet Ryan X-13 Vertijet. These designs took tremendous energy to lift off and land. So a slight variation was incorporated to help get the next designs off the ground, V/STOL: vertical/short take-off and

This Grumman tilt-wing aircraft epitomizes the trend in advanced aeronautical design during the 1960s, showing an executive-type VTOL aircraft in corporate markings of that era. (Northrop Grumman History Center)

Vertical Flight and Other Concepts
Chapter 5

An in-house metal model of the Ryan XV-5 Vertifan. This was an experimental aircraft designed for vertical take-off and landing, commissioned by the US Army. (Author)

landing, or STOVL: short takeoff, vertical landing, describing an aircraft's ability to take off after a very short acceleration run and later (when lower in weight) to land back vertically.

A modern day example of STOVL is the British Harrier (the US Marines' AV-8). Whilst it is perfectly capable of taking off vertically and hovering with a light load, it is incapable of doing so when carrying its maximum weapons and fuel. So it is normally operated with a short take-off run, preferably using a ramp – leaving its spectacular ability to fly straight up, forwards or backwards, strictly for air-shows!

But way before the Harrier ever flew, VTOL and STOVL designs were rigorously explored by a number of countries during the Cold War, and especially by the United States.

This chapter will look at a number of these concepts and projects that struggled on the drawing boards or even as completed prototypes to try to get into the air. Many never made it, nonetheless, there were some successes.

Ryan engineering plans for the XV-5 Vertifan. This drawing reveals the complex plumbing necessary to keep this craft airborne. There seems to be very little room left for weapons. (Scott Lowther)

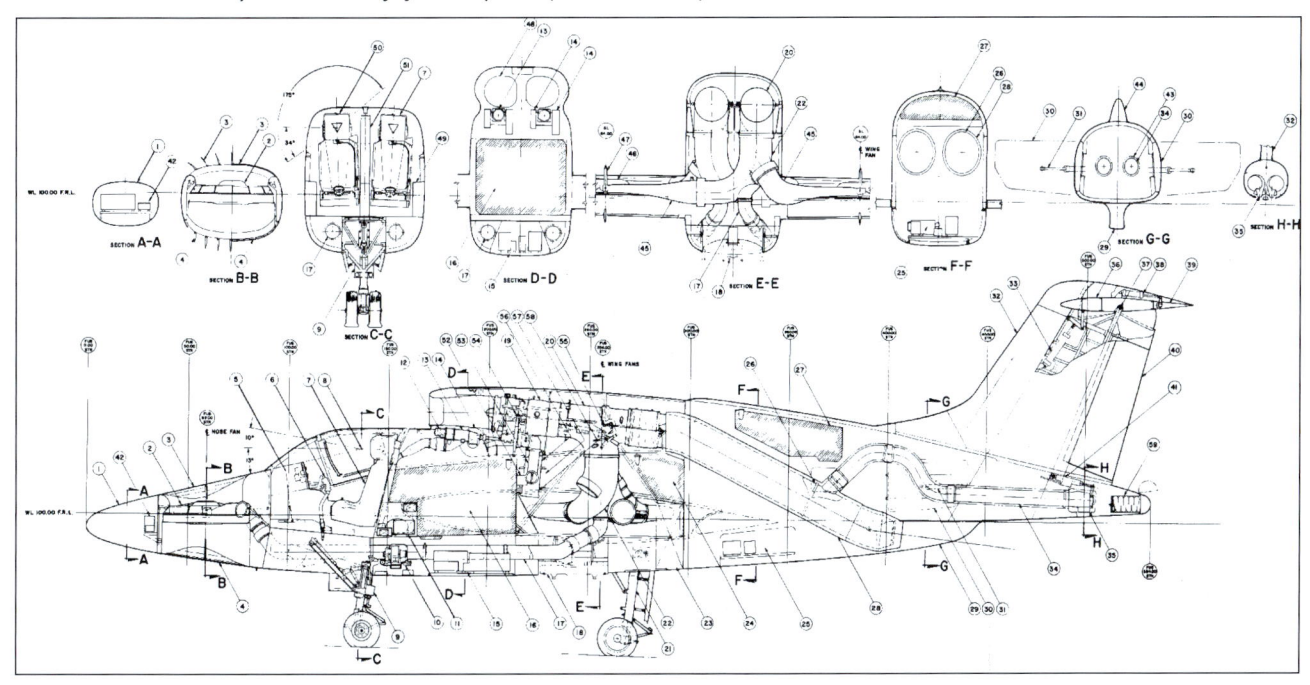

Vectored and Separate Lift Thrust

Although vertical flight had been attempted, and sometimes attained, by various experimental aircraft in the early post-war years, the advent and rapid development of the turbojet engine during the 1950s opened up many new possibilities. They could either drive rotors or lift-fans or have their thrust re-directed downwards. They could even be designed as specialized lift engines, with very high thrust/weight ratios, intended simply to get the aircraft into the air before more conventional engines took over to provide forward propulsion.

Ryan XV-5 Vertifan

The Ryan VZ-11RY was commissioned by the U.S. Army in the early 1960s. It was re-designated the XV-5 Vertifan in 1962, and its first flight was made on 25 May 1964.

The wingspan was 29ft (8.8m) and the length 44ft (13.4m). This experimental aircraft had two J85 turbojets from which the exhaust gases exited through three fans, one in the nose and one in each wing. The wing fans had large covers on the top, divided in halves that flipped up to open for vertical flight. Louvered vanes beneath each fan controlled the direction of air and the flight of the aircraft.

Two VX-5As were built. There was some success in flying, but not all the bugs were worked out. One crashed in 1965 during a public demonstration, and then the other was badly damaged in 1966 during a rescue demonstration, and the pilots were killed in both crashes. The second craft was rebuilt and given to NASA as the XV-5B, and continued flight testing until 1971.

Lockheed XV-4 Hummingbird

At the same time the Army ordered the Ryan, it also asked Lockheed to participate. Lockheed's VZ-10 was re-designated the XV-4 Hummingbird and two were built. This is an example of separate thrust and lift. The Hummingbird had a wingspan of 26ft (7.9m) and a length of 32ft (9.8m). It was powered by two Pratt & Whitney JT12A turbojets.

The Hummingbird had long doors in the upper and lower fuselage to feed air into the lift engines and exhaust gasses out the bottom. The prototype crashed on 10 June 1964, killing its pilot. The second Hummingbird was converted to jet thrust only, and after a number of test flights, it too crashed.

In-house metal model of the Ryan XV-5 Vertifan shown in US Air Force markings. This iteration has a larger and different style wing, fitted with leading-edge external fuel tanks. The horizontal stabilizer is also different than the Army version. (Author)

Vertical Flight and Other Concepts   Chapter 5

In-house metal model of the Lockheed XV-4 Hummingbird. This aircraft has the engines located above the wings. The upper fuselage contained long doors that would open to reveal the lift fans. (Greg Barbiera collection)

This resin model of the Lockheed XV-4 Hummingbird is an Air Force version of the same aircraft with slight variations to the tail structure, wingtip fuel tank shapes, and fuselage openings for the lift fans. (Greg Barbiera collection)

North American V/STOL

North American Aviation developed and proposed their own V/STOL concepts including a tilt-wing transport which is shown later on in this chapter. One proposed vertical trainer derives from the airframe of the Navy's T-2 Buckeye, using modified wings to hold lift fans. Another one, the Sabre V/S bomber proposal, uses vectored thrust. Unfortunately, very little information has yet been uncovered on these models.

In-house North American metal proposal model looks to be a modified T-2 Buckeye. This model may well be the more formal presentation of the one shown on page 153. Inscription on the base of the stand identifies this model as a Class VTB, Trainer System 241. (Author)

151

Chapter 5 Secret US Proposals of the Cold War

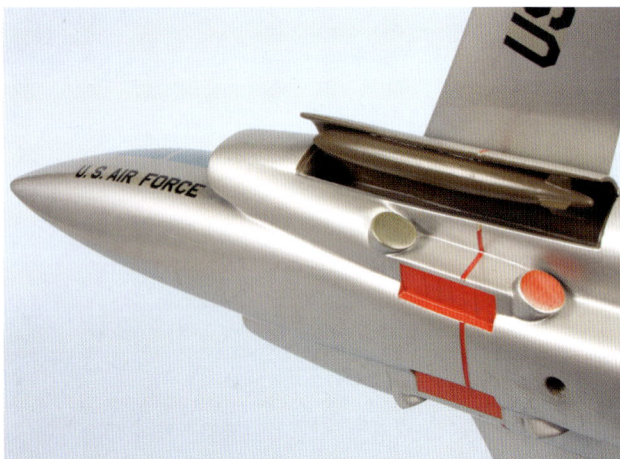

Above: The North American Sabre V/S was a proposal for a V/STOL bomber, but very little information is known about this aircraft. (John Aldaz collection)

Left: A photo taken from underneath the Sabre V/S shows the small, color coded engine exhaust nozzles and vents. These colors match the coloring of the engines seen in the next picture to indicate from where the thrust originates. Also the bomb bay has been opened to reveal its ordinance. (John Aldaz collection)

Sabre V/S

North American came up with a concept for a V/STOL bomber that uses vectored thrust nozzles. This model may have been just an in-house design study and little information has been uncovered as of yet. The model is quite beautiful in its form and has a number of unique features. The thrust nozzles and vents are color coded to match the coloring of the internal engines. There is a transparent section in the upper fuselage that allows the viewer to see the inner layouts of the turbojets and ducting. And the concept even displays a fuselage bay that opens to reveal its bomb.

The Sabre V/S has a see-through fuselage section displaying the turbojets and ducting layout. The front turbine is in the same olive drab color as the two forward nozzles underneath the craft, while the red ducting matches the color of the rear nozzles and vents. (John Aldaz collection)

Vertical Flight and Other Concepts — Chapter 5

V/STOL Evaluation Vehicle T2J Trainer

A common technique to quickly make-up a concept was to modify an existing model. North American model shop took a TJ2-1 Buckeye Topping model, altered it, added a fan lift wing and Army markings, glued on a new plaque and came up with this V/STOL presentation.

Above: *The stand of this converted Topping Model of a former Buckeye jet trainer reads: "V/STOL, Evaluation Vehicle, NAA T2-J Trainer with G. E. Lift Fan System. (John Aldaz collection)*

Grumman Design 260-4C

Grumman also had an Army proposal for a fan-lift surveillance aircraft, the Design 260-4C from 1960. Very little information is known about this proposal, but the model makes an impressive statement of speed and sleekness.

A photo of the underside of Grumman's Design 260-4C reveals large louvered panels in the center inboard section of the wing. (Northrop Grumman History Center)

This beautiful model of the Grumman Design 260-4C shows a proposal for a Lift Fan Surveillance aircraft for the US Army. (Northrop Grumman History Center)

Chapter 5   Secret US Proposals of the Cold War

Republic VTOL Proposals

At the Cradle of Aviation Museum archives and at the American Airpower Museum located in the original Republic hangars in Long Island, a number of unknown Republic vertical flight designs were examined. Among them, four different concepts were picked out to show the different propulsion systems that Republic explored in their quest for vertical flight. These systems include the small lift jets, lift fans, and vectored-thrust to raise their aircraft into the air. One concept has an unusual delta wing with an auxiliary variable-geometry swept wing. Best guess on the dates of these models would be early to mid 1950s.

Shot from underneath, this in-house wood and metal model of Republic's unknown VTOL aircraft reveals six lift jets that are paired and articulate from vertical to forward thrust. (Cradle of Aviation Museum)

Top view of Republic's VTOL concept with the unique pop-up lift jets located in the center of each wing. Also different about this concept is that it is actually a tail-dragger. (Cradle of Aviation Museum)

Above: *An underside view of another Republic unknown VTOL concept showing the three lift fan exhaust ports in the center mid-section of the fuselage. (Cradle of Aviation Museum)*

Below: *Top view of unknown Republic's VTOL aircraft shows highly swept-back wings with the three lift fans in the fuselage. This single-seater may have been an early study of the proposed AP-100. (Cradle of Aviation Museum)*

Vertical Flight and Other Concepts — Chapter 5

Right: *This small delicate in-house wood model of an unknown Republic V/STOL swing-wing design is highly unusual. Very few, if any vertical proposals offered a delta wing with variable-geometry. Here the model is shown with the sesqui-wings fully swept-back.* (Cradle of Aviation Museum)

Below left: *While the variable sweep sesqui-wings are fully extended, the aircraft still retains its delta-shaped main wing. The extended sesqui-wings probably functioned as the horizontal stabilizer and elevator when flying at slow speeds.* (Cradle of Aviation Museum)

Below right: *The underside of this unusual swing-wing V/STOL shows the four louvered nozzles of its lift and thrust propulsion system reminiscent of the British Harrier, but without the articulation capability of its nozzles.* (Cradle of Aviation Museum)

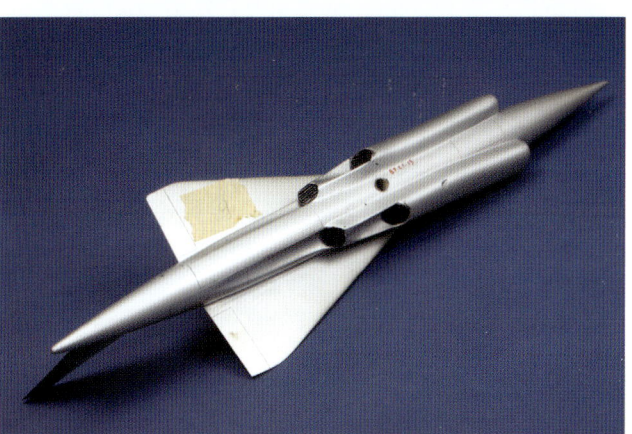

Variable-Geometry Sesqui-Wing V/STOL

This unknown Republic V/STOL concept wood model uses four louvered vectored thrust nozzles along the bottom of the fuselage. But the more unusual aspect is its delta-style wing platform that holds variable-geometry sesqui-wings. When in the forward position, the small wings reveal a very narrow cord and the delta wing platform does not change its shape. No information has yet been found on this design.

Republic seemed to rely on their successful Thunderstreak airframe to create just about any variation possible. Here the basic F-84F has been given overwing engines to power a new V/STOL fighter called the F-84F/V. (George Cox)

F-84F/V

The F-84F/V was a Republic proposal for a V/STOL aircraft using the well-proven F-84 airframe. The engines are located on the wings and supply air to the nozzles located along the centerline, bottom of the fuselage. There are also control jets at the nose, tail, and wingtips for stabilization. Republic spent some effort with this proposal, going to the trouble of producing a model and color illustrations for presentation.

A photo of the underside of the wood F-84F/V model shows the layout for the fan openings aligned along the centerline of the fuselage. (George Cox)

155

Tilt Rotor

While the practical lifting capabilities of large rotor blades were proven as early as 1939 with Igor Sikorsky's novel VS-300 helicopter prototype, the idea of tilting the entire rotor disc to provide both vertical and horizontal flight did not reach fruition until the 1950s. Several of the designs are represented in the following pages. When the enhanced thrust of turboprop engines became available this led to the makings of some very advanced but complex VTOL aircraft.

Douglas Doak

Edmund Doak was interested in VTOL aircraft since the late 1930s and started his own company, the Doak Aircraft Company in Torrance California. As a small, independent manufacturing company, he produced the Doak VZ-4 (Model 16) for the Army, a small, shrouded tilt-rotor aircraft that first flew on 23 February 1958. Eventually Doak went on to work for Douglas Aircraft and continued designing shrouded tilt rotor projects. This unknown Douglas concept is one of his designs, a small transport using a combination lift and thrust propulsion system.

After Edmund Doak went to work for Douglas he designed various VTOL concepts. This is one of the more unusual projects showing a transport aircraft with three small jet pods above the cockpit and shrouded tilt rotors on both wingtips. (Author)

Bell V-Tail/Tilt Rotor

Although outside of our book's timeline this early Bell tilt-rotor concept is included here to offer a quick peek into the future. This in-house study is an early V-tail version of the V-22 Osprey. It would become the only successful tilt-rotor VTOL of the future.

Tilt Wing

Taking the tilt rotor concept to the next level, aerodynamicists and aeronautical engineers conceived V/STOL flying machines where the entire wing would pivot 90° from horizontal to vertical and back again. Once it reached a suitable height, the wing would transition back to horizontal position, to fly normally. Here are some models of those efforts that looked highly plausible, but never left the drawing board.

Grumman Gadfly

Starting in the 1950s Grumman offered a number of tilt-wing proposals that carried on into the 1970s. Some of these proposals were dressed in civilian colors and others in military. The Gadfly was one of the first, and it came in two- and four-engine versions. Grumman also took its successful OV-1 Mohawk observation aircraft and converted it into a tilt wing.

From 1960 comes this corporate four-engine version of the Gadfly, Design 235-7 named the "Grumman Grouse". This concept was designed by Mazzitelli and Skinner. (Northrop Grumman History Center)

Bell made a number of studies for their tilt-rotor project and this in-house resin model seems to be one of a proposed V-tail aircraft. (Author)

In the early 1960s Grumman came up with Model 134E, a proposal to add a tilt-wing with four turboprops and perforated flaps to an extended OV-1 Mohawk fuselage. Note the horizontal tail rotor used for pitch control. (Northrop Grumman History Center)

A tilt wing aircraft called the Helicat is seen dressed up in civilian colors. This concept was created by Grumman designers Steinbach and Kirschbaum and progressed at least as far as this wood model. (Northrop Grumman History Center)

Chapter 5 — Secret US Proposals of the Cold War

An in-house wood model of Grumman's Design 242 study for a tilt-wing assault transport. This is a very impressive model with its four-engine tilt wing and twin tail. It is dated 1959. (Northrop Grumman History Center)

A top view of Grumman's Design 242, an assault transport concept showing the tilt-wing in the raised position for short take-off into vertical flight. Note the shrouded tail rotor and the rear ramp door in the down position. (Northrop Grumman History Center)

A four-bladed version of North American Tri-Service tilt-wing transport proposal. The V/STOL was purported to be able to lift a 4,000-pound (1814 kg) payload with a maximum speed of 240 knots (444.5km/h). (John Aldaz collection)

North American Tri-Service V/STOL

The Tri-Service V/STOL was a tilt-wing 1960s proposal with four 500hp (372.9kW) turboprop engines driving four 13ft (3.9m), three-blade propellers, and a tail-mounted fan for pitch control and added lift. Its mission was to transport troops or equipment up to 4,000lb (1814.4kg) in weight. It had a maximum radius of 250nm (463.km) at a cruise speed of 200 knots (371km/h) without refueling. Takeoff gross weight for VTOL was 11,000lb (4,990kg); for STOL it was 14,000lb (6,350kg). Maximum airspeed was 240 knots (444.5km/h).

Unknown Tilt-Wing Design

This large four engine, tilt-wing model of a heavy transport was found tied to the rafters of an old storage facility. It is thought to be made by Fairchild-Hiller, somewhere around 1965. It has an unusual camouflage scheme and a linkage that connects the wing to the elevator. As the wing is tilted up or down the elevator moves along with it for nose-up compensation.

Not much is known of this large four-engine tilt-wing transport. Speculation is that it comes from Fairchild-Hiller. The model is made from wood and metal and is quite heavy. It has a wingspan a little over 4ft (1.2m). (Author)

Helicopters

With roots of rotary-wing designs dating back to primitive contraptions first sketched by Leonardo da Vinci during the Renaissance, the practical evolution of rotary-wing flying machines began in several different countries around the world in the late 1920s. By the late 1930s, the possibility of piston-powered helicopters was becoming a realistic concept.

Sikorsky

The father of the American helicopter was a Russian-born engineer and designer, Igor Sikorsky. He was the first to design and build a stable, fully controllable single-rotor helicopter, the VS-300, an evolution of which went into production in 1942. Several of these early aircraft saw limited action as rescue platforms in the final days of World War II, although the helicopter did not come into its own until the Korean War eight years later. The Sikorsky Aircraft Company, to this day, remains one of the world's leading manufacturers of large military and civilian helicopters.

A Sikorsky Engineering plan of the DS-160 shows it configured for carrying men and cargo. It also shows the attached "airborne trailer" slung under the fuselage, the basic principle of the company's large S-64 Skycrane cargo helicopter that flew decades later. (Scott Lowther)

DS-160 Utility Helicopter

Offered in 1945, the twin-engine "Flying Crane" was proposed as a large transport. Its power came from two Wright/Lycoming R-1300 engines and the craft had a gross weight of 16,423lb (7,449.3kg). The fuselage could hold 21 men and in the "airborne trailer," carried under the fuselage, an additional 21 men. The trailer could also hold a combination of men and equipment. The DS-160 could be configured with either landing wheels or pontoons.

Without its double rotors this Sikorsky DS-160 looks rather sparse. This is all that is left of this original in-house presentation wood model. Sikorsky offered a double rotor or a single rotor version, but neither were ever built. Note external engine pods. (Larry McLaughlin collection)

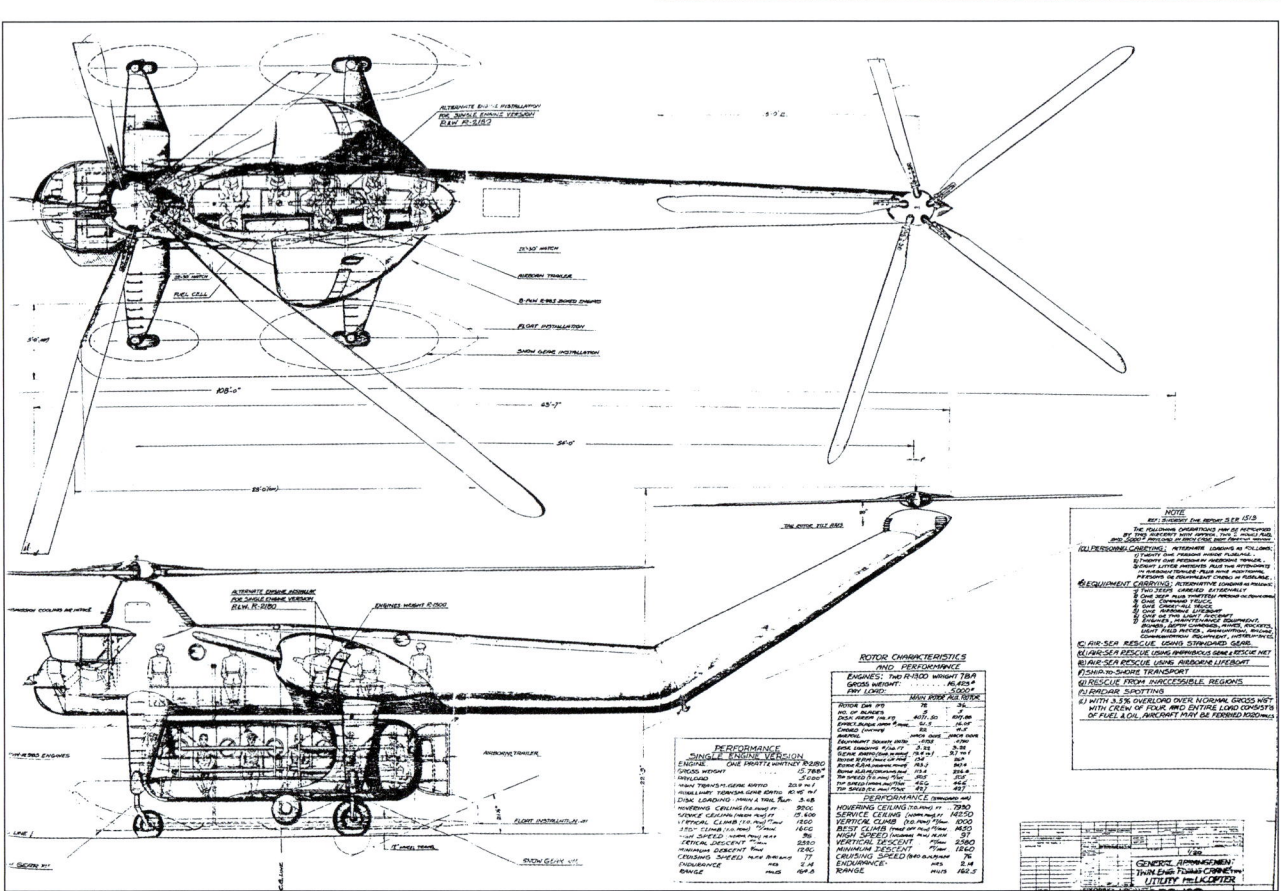

Vertical Flight and Other Concepts

Chapter 5

Above: *This in-house model of Sikorsky XHSPA-1 survived without its rotors but it is still a very impressive piece. The interior of the cockpit is highly detailed.* (Larry McLaughlin collection)

Right: *A bottom view of the Sikorsky XHSPA-1 model reveals the single large bomb bay with its deadly payload. Painted in an early-1950s Gloss Sea Blue US Navy color scheme, this helicopter must have been intended for ASW (anti-submarine warfare) missions.* (Larry McLaughlin collection)

XHSPA-1

More research needs to be done on this XHSPA-1 proposal from Sikorsky. This large and impressive model, over 3ft (1m) long, has long since lost its aluminum main rotors, all that remains is the body. Because of its size, the cockpit has been fully detailed with pilot seats and instruments. Under the fuselage is a bomb bay with a transparent door that reveals the large bomb it carries.

Piasecki HRP-1

If one ever wondered where the nickname "The Flying Banana" came from, here it is. Piasecki's original proposal for a twin-rotor helicopter was known as the HRP-1 Rescuer or the "Harp." It first flew on 7 March 1945. The HPR-1 became the first U.S. military helicopter with a significant transport capability, and was used for cargo and personnel. It was powered by the 600hp (447.4kW) Pratt & Whitney R-1340-AN-1 engine.

Twin-rotor Piasecki HRP-1 was the original concept for the "Flying Banana" and the aircraft was the progenitor of today's massive CH-47 Chinook helicopter. The in-house model has a wooden fuselage and aluminum rotors. (Larry McLaughlin collection)

Hughes XH-28 Heavy Lift

In 1946, the U.S. Army Air Force awarded Kellett Aircraft a contract to design a heavy-lift helicopter, the XH-17. Hughes Aircraft took over the program in 1948, and the helicopter finally flew in October 1952. The XH-17's 130ft (39.6m) diameter two-blade rotor was the largest ever flown, with each blade powered by tip jets. Kerosene burners at the tip of each blade were fed by two General Electric TG-180 turbine engines acting as compressors. Although the XH-17 could lift more than 4.5 tons (4082.3kg), it required too much fuel to have any effective range. It was also plagued by rotor vibration and blade fatigue.

In the early 1950s, the XH-28, a newer and bigger design, was proposed and construction started. It was designed to lift 18 tons (16,329kg), but construction was stopped because of the Korean War.

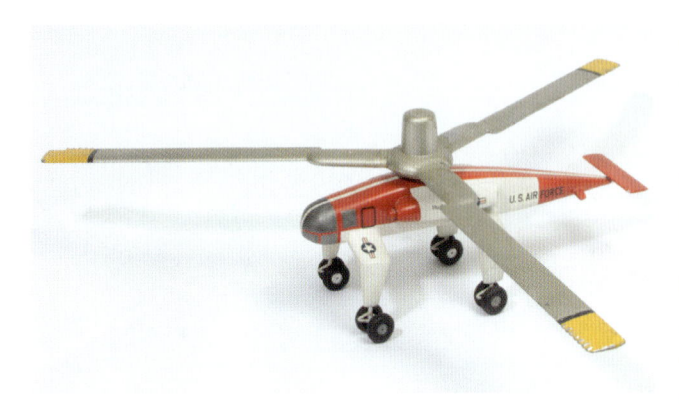

Although the XH-28 was to be made with a four-blade rotor, this model shows a three-blade rotor. At the end of each blade are indications of small jets used for direct hot-cycle thrust. To give an idea of scale, two large trucks would be able to fit side-by-side between the wheels of the XH-28. (Author)

The unusual Hughes concept had a large Y-shaped rotor/wing that would be stopped in flight and locked into position so that two of the blades acted as wings. The idea was to be able to take-off and land like a helicopter, and fly like a jet. This aircraft was to be a rescue or transport vehicle. (Blake Sutton)

Hughes Y-Wing

In the 1960s, Hughes was interested in designing a helicopter that could behave as a helicopter for vertical landings and takeoffs but, once in the air, turn into a jet. Its method was to use a hot-cycle reaction drive to power a Y-shaped rotor/wing. It worked as a rotor for vertical lift, could be stopped and locked into place during flight, and then function as a wing for high-speed forward flight. This reduced drag and increased high-speed efficiency.

The hot cycle system was the same used in the XH-17 and the proposed XH-28, where hot gasses were sent into a collective hub, channeled through the rotors, and out the tip jets positioned on the end of each blade. Wind tunnel tests stopped the project when it was discovered that the large rotor caused vibration problems.

Grumman and Lockheed Stopped/Stowed S/VTOL

Two other helicopters employed a slightly different approach to the Hughes Y–Wing concept. These designs not only stopped the rotors in flight, they actually stowed them away once the aircraft transitioned into high-speed flight. Very little information has been found on the Lockheed design, and only the model of the Grumman exists as any evidence that this concept was even being considered.

An in-house model of a Grumman proposal for a folding-rotor helicopter/airplane hybrid. With its rotor in the stowed position as seen here, the airframe of this machine resembles one of today's fast attack helicopters. (Larry McLaughlin collection)

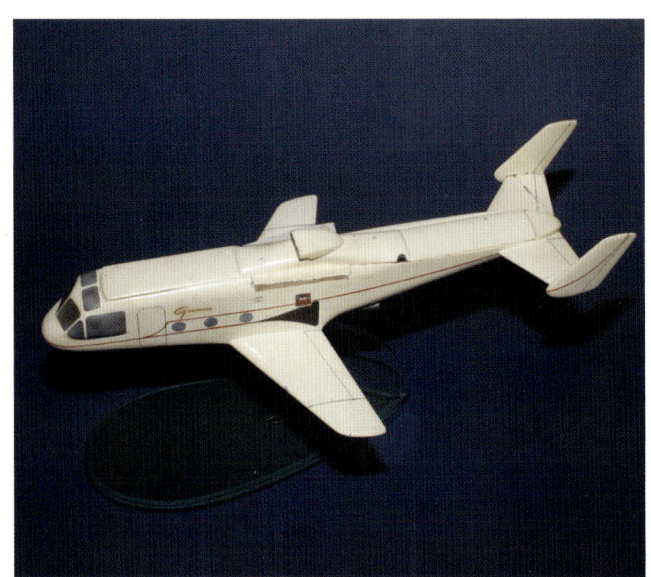

Vertical Flight and Other Concepts　　　　　　　　　　　　　　　　　　　　　　　　　　　　　　　　　Chapter 5

Above top: *Grumman's folding rotor proposal model is shown with its rotor in the extended position as it would be for flight as a helicopter. Note the large overhead "eyebrow" windows above the cockpit for enhanced pilot visibility.* (Larry McLaughlin collection)

Above: *This highly detailed Lockheed helicopter concept model is shown with its rotor deployed for flying like a fast helicopter.* (John Aldaz collection)

Right: *Here, the Lockheed helicopter concept model has its rotor blades in the stowed position ready to fly as a fixed-wing aircraft.* (John Aldaz collection)

163

Drones

During the Cold War, drones had varied meanings and uses: they were unmanned aircraft put up for target practice; Navy Loons were copies of German V-1s used to test guided missiles; and other drones were employed for surveillance.

Since the turn of this last century drones have gone through a dramatic growth. Recent warfare in the Middle East has suddenly shifted the development of drones into high gear. And with the benefit of today's technologies these UAVs (uninhabited aerial vehicles) come in many different forms, shapes and sizes. Now they are called ISRs: Intelligence, Surveillance and Reconnaissance. Here is a look at some of the early Cold War drones.

Globe Model 22 "The Goblin"

Globe Aircraft developed target drones for the U.S. military after WW II utilizing piston and pulse-jet power systems. Their first pulse-jet drone, the 1946 XKD2G-2 Firefly, used McDonnell's 8in (20.3cm) PJ42 engine. By 1950 the KD2G-2 operated a Solar PJ32, the XKD5G-1 flew on a Marquardt PJ46. At one point Globe Aircraft proposed a very German looking drone, the Globe Model 22, named "The Goblin." It had a delta-wing similar to the Lippisch DM-1 attached to what appears to be the Ford PJ31, a copy of the German V-1's Argus pulse jet.

It is not known whether Globe consulted with Lippisch, as he was for Convair's XF-92A/F-102 delta-wing program (Chapter Three), or if his World War II research on delta wings was used. This information was readily available to U.S. companies through NACA.

Globe Aircraft's Model 22 "Goblin" shows the German influence from WWII. This model of a target drone has a modified Alexander Lippisch's delta wing from his DM-1 glider and is powered by a copy of the German V-1 Argus pulsejet. (John Aldaz collection)

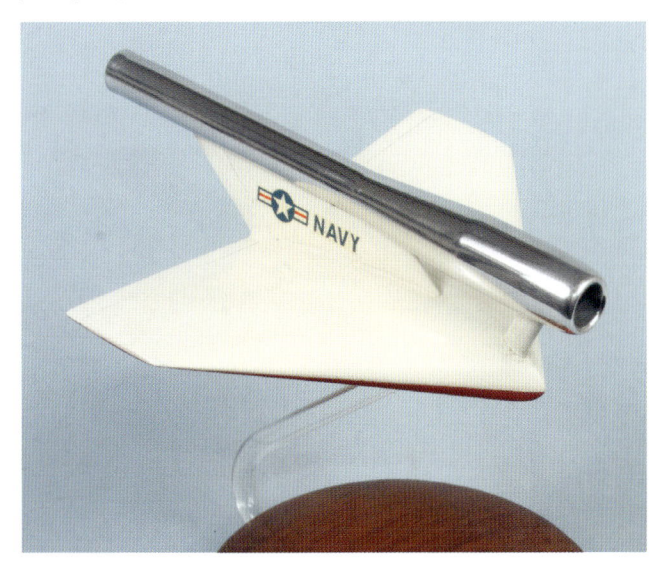

Grumman Design 167

Grumman's VTOL drone Design 167 from May 1957 has a tilt-wing with rotors for vertical lift and forward flight. This appears to be the presentation model for the Army's consideration, but little data has been found on this proposal.

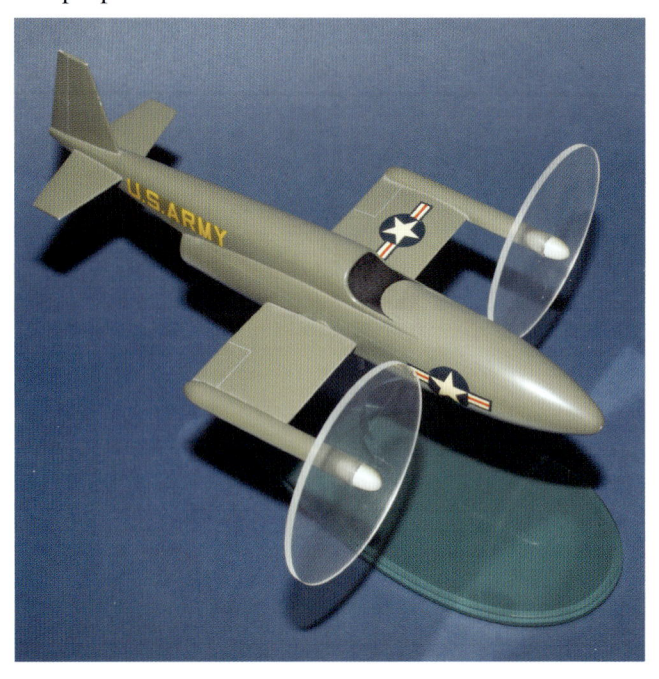

Grumman Design 167 VTOL drone has its tilt-wings in the forward thrust position. Note the large air inlet located at the top mid-section of the fuselage. Power source is either a turbojet or could be a turboshaft. (Larry McLaughlin collection)

In-house wood proposal of the Design 167 drone shown with its tilt-wing and rotors in the vertical flight position. The use of plexiglas disks instead of static propellers improve the appeal of the model for they give it the dynamic sense of flight. (Larry McLaughlin collection)

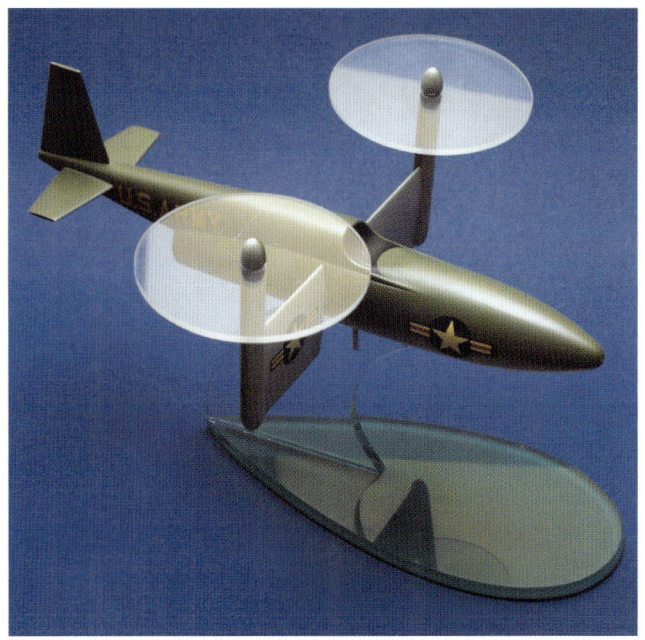

Convair LALO and ADD

Convair began conducting tests and studies in 1959 of remote-controlled ducted-rotor systems with VTOL capabilities. The Low Altitude Observation System (LALO) was a drone developed for the Army. Think of it as the modern version of the American Civil War hot air balloon, where a soldier was lifted up a couple hundred feet and shouted down to the generals what was happening on the battlefield. But in this case, LALO could go as high as 10,000ft (3,050m) and send back reports in real-time video using a TV camera. It had a range of 3 miles (4.8km), and an endurance of one hour of flight time. This diminutive drone was only 4ft (1.22m) in diameter.

The ADD, or Advanced Destroyer Drone, was a larger version of the LALO. It was more than 10ft (3.05m) high, with an 8ft (2.4m) diameter ducted fan. Powered by a 650hp (484.7kW) engine, this aircraft had the ability to carry two Mk 44 or Mk 46 torpedoes.

Right: *In-house model of the Convair ADD showing how the drone looks in its assembled form. The circular wing design was very similar to the French "Coléoptère" VTOL testbed.* (John Aldaz collection)

Below: *Convair's model of the Convair ADD (Advanced Destroyer Drone) is extremely well crafted and can be disassembled to reveal the complex interior structure of the aircraft.* (John Aldaz collection)

Launched from a ship, the drone could seek out enemy submarines and drop depth charges or homing torpedoes. Interestingly enough, this technology was revived again in the 1980s and it is rumored that development continues to the present time.

Convair Model 49

The Model 49 can be considered a helicopter and should be included in the helicopter section of this chapter. However, it is placed here, as it seems to be the evolution of the LALO and the ADD drones mentioned above.

In 1965, the U.S. Army called for submissions of an Advanced Aerial Fire Support System (AAFSS). It was to provide cover for ground troops during firefights. Most aerospace companies responded with advanced helicopter designs, but Convair came up with something truly different: Essentially, the proposed Model 49 was a "flying tank;" a circular, closed-wing vehicle almost 23ft (7m) in diameter with an articulated cockpit that was 30ft 2in (9.2m) high. It had two three-blade contra-rotating rotors powered by three Lycoming LTC 48-11 engines. It was also armed with heavy on-board weaponry.

Similar to a helicopter, the Model 49 could take off and land vertically. And it would fly horizontally like an airplane. The nose could articulate downwards, 45-degrees so the pilots could oversee landings or take-offs or to direct their weapons in battle.

In-house model of the Convair Model 49. This was a larger, manned version of the ADD. It was intended to be a two-man flying tank capable of taking off and landing vertically. The canopy would be articulated to allow for landing and taking off with the cockpit at a 45-degree angle while resting on the ground. (Author)

Conclusion

The aircraft and projects described in this book illustrate the progress of technology and the history of aircraft design over a twenty year period. It was a time during which not only were great advances made but new possibilities opened up.

However, it was also an era when the demands of the military and the ambitions of the aircraft manufacturers often outstripped the knowledge and capability of the day. Many of the proposed designs could not be turned into practical operational aircraft. Some never left the drawing board; others progressed as far as full-scale mock-ups; a few even got as far as flying prototypes before their shortcomings were revealed. However, in many cases the projects were simply ahead of their time. It would take many years before the concept would reach fruition. The flying wing bombers of the late 1940s and early 1950s would lead eventually to the remarkable B-2 "Stealth Bomber" – but it would take half-a-century.

Perhaps nothing illustrates this long-term pursuit of a goal more than the quest for the vertical take-off and landing fighter, particularly one that could compete on all accounts with the equivalent high-performance conventional aircraft. Many different approaches to this concept were explored in the 1950s and 1960s, but only one – the British use of a single jet engine with deflected exhaust – led to an operational warplane. The Soviets put a VSTOL aircraft, the Yak-38 into service with their Navy but it was more of a short-lived trial than a genuine operational deployment. The Harrier on the other hand was to serve with several air-forces and navies around the world, including the US Marines (under the designation AV-8). It would also prove itself in several conflicts both as a ground attack aircraft and an interceptor. However, even the Harrier did not meet the goal of being capable of level supersonic flight.

This capability has had to wait until today and what is arguably the most advanced aircraft in the world: the Lockheed Martin F-35 Lightning II. With its vectored-thrust Pratt & Whitney F135 main engine coupled with a Rolls-Royce Lift Fan, the F-35B not only lands and takes off vertically, but also transitions to level supersonic flight. When it finally enters service with the Western world's major air arms later in this decade, the F-35B will be the world's first and only operational supersonic VTOL stealth aircraft. That is more than sixty years after the requirement was first envisaged by the military.

Many books have documented aspects of aircraft development over this period. They have usually concentrated on particular aircraft types or the history of various manufacturers. The notable exception has been Tony Buttler's remarkable *Secret Projects* series dealing with the design proposals of different nations, many of which have never been revealed before. Much of the information has been lost or destroyed, though patient searching through archives continues to shed light on this fascinating era. However, surely nothing brings this history more to life than the models which the manufacturers themselves used to communicate and demonstrate their ideas?

One can look at one of these models today, as illustrated in the pages of this book, and easily imagine the excited designers, aircraft company executives, airforce generals or navy admirals gathered round inspecting it. Occasionally, one even comes across a picture of such an occasion, with one of the surviving models – the very same item – being examined by a famous designer or test pilot.

Many of these models have sadly been destroyed over the years, thrown away once the project ended. A number have, however, survived: some in museums, some in the hands of private collectors. Hence the ability to produce this book. Every so often, another model will turn up, perhaps in the estate of a retired engineer or aircraft company employee. Occasionally it might even represent a hitherto unrecorded project, a missing link in the design chain. Each time this happens, if it falls into knowledgeable hands, it will be lovingly cleaned and restored – and a further little piece of aviation history will have been preserved.

Almost 70 years after the XP-59 flew over Muroc as depicted on Page 86 this aircraft is banking over the testing grounds now known as Edwards Air Force Base representing the ultimate benefactor of all the technologies, tried, tested, abandoned and advanced. That airplane is the Lockheed Martin X-35. (Mike Machat collection)

Bibliography

Anderson, Fred, *Northrop: An Aeronautical History*, Northrop Corporation, (1976)

Bradley, Robert E, *Convair Advanced Designs: Secret Projects from San Diego, 1923-1962*, Specialty Press, (March 15, 2010)

Buttler, Tony, *American Secret Projects: Bombers, Attack and Anti-Submarine Aircraft – 1945 to 1974*, Midland Publishing Ltd, (October 21, 2010)

Buttler, Tony, *American Secret Projects: Fighters & Interceptors 1945-1978*, Midland Publishing Ltd, (April 15, 2008)

Huenecke, Klaus, *Modern Combat Aircraft Design*, Naval Institute Press

Jenkins, Dennis R and Don Pyeatt, *Cold War Peacemaker: The Story of Cowtown and the Convair B-36*, Specialty Press, (January 15, 2010)

Jenkins, Dennis R and Landis Tony R, *Experimental & Prototype U.S. Air Force Jet Fighters*, Specialty Press, (April 15, 2008)

Jenkins, Dennis R and Landis, Tony R, *Valkyrie: North American's Mach 3 Superbomber*, Specialty Press, (January 15, 2005)

Jenkins, Dennis R, *Space Shuttle: The History of the National Space Transportation System the First 100 Missions, 3rd Edition* Dennis Jenkins, (May 11, 2001)

Landis, Tony, *Lockheed Blackbird Family: A-12, YF-12, D-21/M-21 & SR-71 Photo Scrapbook*, Specialty Press, (February 1, 2010)

Lawson, Robert L, *The History Of US Naval Air Power*, The Military Press, (May 15, 1987)

Lowther, Scott, *Aerospace Projects Review*, Volume 1, Number 4, July-Aug 1999

Lowther, Scott, *Aerospace Projects Review*, Volume 2, Number 1, Jan-Feb 2000

Lowther, Scott, *Aerospace Projects Review*, Volume 2, Number 3, May-June 1999

Lowther, Scott, *Aerospace Projects Review*, Volume 4, Number 5, Sept-Oct 2002

Lowther, Scott, *Aerospace Projects Review*, Volume 5, Number 4, Jan-Feb 2003

Lowther, Scott, *Aerospace Projects Review*, Volume 5, Number 5, Sept-Oct 2003

Lowther, Scott, *Aerospace Projects Review*, Volume 5, Number 6, Nov-Dec 2003

Maloney, Edward T, *Northrop Flying Wings*, Aviation Book Company, (August 1980)

Norton, William, *U.S. Experimental & Prototype Aircraft Projects: Fighters 1939-1945*, Specialty Press, (September 1, 2008)

Thomason, Tommy H, *Strike from the Sea: U.S. Navy Attack Aircraft from Skyraider to Super Hornet, 1948-Present*, Specialty Press (July 15, 2009)

Thomason, Tommy H, *U.S. Naval Air Superiority: Development of Shipborne Jet Fighters - 1943-1962*, Specialty Press, (February 15, 2008)

Whitford, Ray, *Design for Air Combat*, Jane's Information Group, (April 1987)

Winchester, Jim, *Concept Aircraft: Prototypes, X-Planes, and Experimental Aircraft*, Thunder Bay Press, (October 7, 2005)

Yenne, Bill, *Convair Deltas: From Sea Dart to Hustler*, Specialty Press (August 15, 2009)

Zichek, Jared A, *Secret Aerospace Projects of the U.S. Navy: The Incredible Attack Aircraft of the USS United States, 1948-1949*, Schiffer Publishing, Ltd, (January 28, 2009)

Zichek, Jared A, *Mother Ships, Parasites & More: Selected USAF Bombers, XC Heavy Transport and FICON Studies, 1945 – 1954*, Jared A. Zichek, (2011)

Index

A
Air Trails 64, 70
Aircraft Nuclear Propulsion (ANP)......64, 172
Allison..................................21, 126, 133
    T40-A-6 ....45, 133, 134
    V-1710-133 ....18, 93
    XT 40 ....40, 41
    XT 44 137
Army Air Corps 18, 38, 39, 40, 93, 131
Army Air Forces 47, 50, 55, 64, 131
Aubin, Brian 8, 13

B
Beaumont, Arthur C.............................19, 25
Bell.....................................33, 78, 98, 156
    P-63 King Cobra 107
    Rascal...60, 62
    X-1 ....73, 103, 106, 143
    X-5 34
    XP-59 ....64, 86, 87, 88, 126
Boeing...................17, 47, 50, 61, 64, 76, 78
    707 85
    B-17 80
    B-29 ....22, 80
    B-47 47
    Model 424 53
    Model 432 ....53, 54
    Model 462 ....51, 52, 53, 54
    Model 724 75
    Model 2707 ....83, 84
    MX-1712 ....73, 74

Buttler, Tony 7, 12, 115, 120, 167, 169

C
Consolidated Vultee 36, 43, 50
    B-36 10
    Model 112 50, 51, 65, 98
Convair 36, 43, 47, 50, 52, 54, 56, 66, 67,
        73, 78, 79, 80, 85, 110, 133, 144,
        146, 148, 165, 169
    B-36 60
    B-60 37
    Model 23A 147
    Model 49 166
    MX-1964 75
    NB-36H 65
    NX-2 ....67, 68, 69
    Sea Dart 145
    XA-44 ....50, 51
    XAB-1 72
    XB-36 ....36, 37, 38, 47
    XB-60 64

XF-92/XF-92A 34, 35, 59, 60, 61,
        62, 106, 107, 164
XP-81 ....90, 91, 128
Crossfield, Scott...73
Cruver 26, 27, 28, 29, 122, 123
Curtiss-Wright.....................98, 99, 103, 111
    J67-W3 110
    P-304-04 ....93, 96, 97
    P-304-05 ....93, 96, 97
    XP-55 Ascender 93, 96

D
DiNoia, Bill..8, 13
Douglas 9, 56, 98, 156
    A-26 92
    AD-1 Skyraider.....................................135
    A3D Skywarrior................................62, 136
    A4D-1 Skyhawk 140
    B-42 ....23, 24, 25
    D-558-II Skyrocket 73
    F4D-1 Skyray 62
    Model D-640 140
    Model 499C 139
    Model 594A 139
    Model 1018 45
    Model 1112 48
    Model 1155 ....48, 49
    Model 1186 ....139, 140
    Model 1211 ....43, 58, 51, 61
    Model 1240 ....9, 43, 60, 61, 62, 63, 64
    Model 1251-A 63, 64
    Model 1355 115
    Model 2229 85
    OS-115 139
    X-3 ....62, 139, 140
    XA2D Skyshark............126, 133, 134, 135
    XB-42 ....18, 19, 22, 25, 93
    XB-43 43
    XF3D-1 141
    XF4D 59, 60

F
Fairchild 16, 64, 65, 159
Model M-128 ....136, 137
FICON (Fighter Conveyor) 60, 169
Focke-Wulf
    Ta-183 ....123, 125
    190-A8 139

G
General Electric 56, 64, 65, 80, 87, 88
    7E-XJ53-GE-X-25 62
    Direct Air Cycle 67, 68, 69

I-16...128
I-XJ53-GE-X-25 63
J93-GE-3R 119
TG-100......................................91, 127, 128
TG-180..............18, 21, 22, 43, 47, 50, 51,
        93, 97, 162
TG-190..103
X279A 75
XJ73-GE-5 100
General Operational
    Requirements (GOR) 75
Grumman 7, 8, 10, 12, 101, 102, 130,
        141, 148, 156, 162, 163
    D-118 142
    Design 75 ....132, 141
    Design 79-D 141
    Design 97 ....11, 142
    Design 118 141
    Design 167 VTOL 164
    Design 242 158
    Design 260-4C 153
    F11F Tiger...142
    G-98 142
    G-107-3 ....118, 119
    Grumman Model Shop 8, 10, 13,
        16, 17, 111
    Model 134E 157
    XF-101 Jaguar 141
    XF5F-1 Skyrocket 131
    XP-50..131

H
Hess, Allen 3, 4, 7, 14, 15, 32, 35, 39, 40,
        56, 87, 88, 89, 95, 100, 122,
        128, 129, 131, 134, 135, 139
Horten 32, 34

K
Keirn, Donald 64, 72

L
Lippisch, Alexander 34, 35, 51, 107, 164
Lockheed.................17, 80, 84, 88, 119, 121,
        162, 163, 169
    A-12 ....78, 80, 121
    CL-288-1 ....115, 116
    F-104 ....108, 114, 116
    F-35 Lightning II 167
    Helicopter 162, 163
    L-133 89
    L-187-7 137
    L-205 ....106, 108, 109, 110, 114
    L-238 Super Neptune 45, 46

L-227-1 114
P2V-7 45
P-80 ....90, 107
SR-71 ....13, 56, 122
SST 84
U-2 ....56, 60
VZ-10 150
X-35 167
XFV-1 Salmon 133, 148
XV-4 Hummingbird 150, 151
YF-12A 121

M
Martin, Glenn L. 17, 47, 50, 51, 52, 53,
54, 56, 98, 100, 144,
146, 167, 168
McDonnell 164
F2H Banshee 124
F3H Demon 143
XF-85 Goblin 58
Messerschmitt
Me-163 Komet 31, 114
Me-262 Schwalbe 30, 86
Me P.1101 V1 ....33, 34
Moore, Chris 7
MX 108
Boeing 73, 74
Convair 75, 78, 106
656 Douglas 139
Republic 102, 109

N
National Air and Space Museum 7, 22
North American Aviation 8, 16, 47, 80,
81, 85, 90, 99, 151, 169
Advanced Piloted Interceptor 116, 117
F-107 106
F-86D 103
FSW P-51 92
Rapier III 112
WS-202A 117
XA2J Super Savage 133
XF-108 120
North American Rockwell 7, 14, 15, 77
Northrop 7, 8, 10, 17, 32, 34, 38, 39, 41,
42, 59, 69, 72, 88, 93, 119, 169
B-2 42
B-62 Snark 62
F-89 Scorpion 100
P-61 Black Widow 98, 99
YB-49 40
XB-35 ....32, 38, 39, 40
XP-56 Black Bullet 95
XP-89 Scorpion 99
XSSM-A-3 Snark 61
Northrop Grumman 7, 8, 11, 12, 13, 17,
101, 118, 119, 131, 132, 141, 142, 148,
153, 156, 157, 158
Northrop, Jack 38
Nuclear Energy for Propulsion of
Aircraft (NEPA) 64

O
OS-111 ....136, 137, 138
OS-113 143
OS-115 ....136 139

P
Pratt & Whitney 64, 65, 80, 88, 93, 167
Indirect Air Cycle 68, 69, 147
J58 121
JT12A 150
R-1340-AN-1 161
R-2000-7 130
R-2800-29 95
R-4360 ....10, 38, 39, 65
XT34-P-10 62

R
Republic 8, 17, 51, 81, 106, 108
AP-31 ....102, 106
AP-42 52
AP-46 ....133, 134
AP-54 ....109, 110
AP-55 ....112, 113
AP-57/XF-103 ....109, 110
F-84 90
F-84F/V 155
F-91A 104, 105
F-105 103
NP-48 143
NP-49 ....143, 144
NP-50 138
RC-2 82
RF-84 60
V/STOL 155
VTOL 154
XF-91 103
XP-91 104
YF-84F 60, 62
Request For Proposals (RFP)...38, 50, 52, 80

S
Saunders-Roe
SR.A/1 144
The Princess Flying Boat 65, 67, 146
Stereo Lithography Apparatus
(SLA) 14, 17
Sikorsky
DS-160 160
Igor 160
VS-300 156
XHSPA-1 161
Strategic Air Command (SAC) 37, 57

T
Topping Models 16, 153

U
U.S. Air Force, USAF 12, 16, 61, 72, 78,
99, 100, 108, 110, 115, 119,
120, 121,126, 133, 169
USAAF 6, 47, 50, 51, 53, 56, 87, 88, 90,
91, 98, 102, 128, 131
U.S. Army 10, 150, 162, 166
USN 146
U.S. Navy 169
USS *United States* 136, 139, 169

W
Weapon System 65, 100
WS-110A 75
WS-125A 65
WS-202A 115, 116, 117
WS-204A 110
Webb, Barry 38, 94, 95, 107, 121, 129, 130
Westinghouse 92, 107
J34 ....127, 128
J34-W-36 140
J34-WE-7 ....58, 98, 128
J34-WE-17 ....139, 140
J34-WE-36 45
X24C-10 140
XJ34-WE-7 98
XJ40-WE-10 140
XJ40-WE-12 ....136, 149
XJ46-WE-2 ....136, 139, 49
19B-2A 18
24C turbojets 141
Whittle, Sir Frank 64, 87
Whittle engine 64, 87, 88
Wiking 27, 28, 29

Y
Yeager, Chuck 73, 102

Other books from Crécy Publishing

Project Terminated

Famous Military Aircraft Cancellations of the Cold War and What Might Have Been

Erik Simonsen

Aerospace history is a fascinating subject. However, what is genuinely intriguing is an examination of those twists and turns of fate, sometimes referred to as, "what might have been". New aeronautical designs are often developed in response to a particular need for which the government may ask industry for input. Yet, in the many attempts to achieve a viable product even the competition winners do not always survive, and are subsequently cancelled for poor performance, not meeting schedule milestones, budgetary pressures or political intrigue.

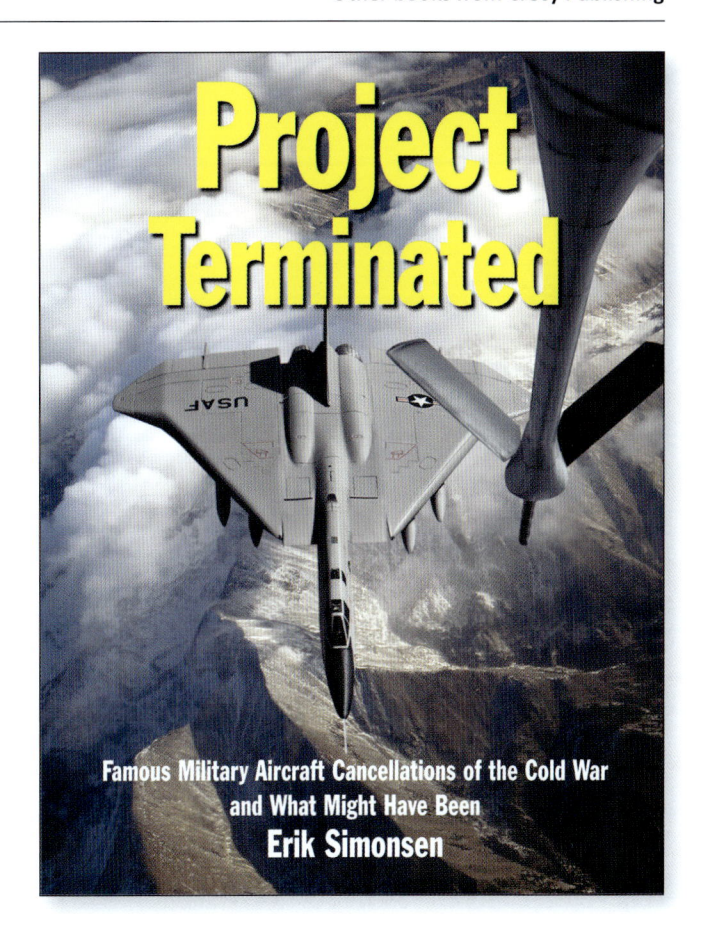

Each aircraft requires very advanced thinking in aerodynamics, materials, manufacturing techniques, training and logistics and where some of these advances were long contemplated by designers such as Jack Northrop and the YB-49, others, such as the Boeing X-20 Dyna-Soar were radically new, forced by the rapid advance of science and warfare. Imposing their will upon the instincts and the experience of their military subordinates in such diverse programs as the North American Rockwell XB-70, Boeing X-20, Lockheed F-12B, Rockwell B-1A, Avro CF-105 Arrow, BAC TSR.2 and the Northrop F-20, politicians often seal the fate of promising contenders.

Project Terminated provides a succinct, accurate assessment of the development of these aircraft, analyzing technical and political challenges and their solutions. Combined with the concept of how these remarkable aircraft would have appeared in operational use, and illustrated throughout with over 250 photographs and drawings, *Project Terminated* provides an enticing look at both the past and the future.

ISBN 9 780859 791731

Binding: Hardback

Dimensions: 290mm x 216mm

Pages: 224

Photos/Illustrations: Over 200 colour and 25 b+w photographs

Price £23.95 $39.95

All titles from Crécy Publishing Ltd,
1a Ringway Trading Est, Shadowmoss Rd, Manchester, M22 5LH.
Tel 0161 499 0024
www.crecy.co.uk

Distributed in the USA by Specialty Press,
39966 Grand Ave, North Branch, MN 55056 USA.
Tel (651) 277-1400 / (800), 895-4585
www.specialtypress.com

Other books from Crécy Publishing

Scooter

The Douglas A-4 Skyhawk Story

Tommy H Thomason

Few modern military aircraft can claim the longevity and overall success enjoyed by the legendary Douglas A-4 Skyhawk. Nicknamed "Heinemann's Hotrod," "Bantam Bomber," and "Scooter," the small, nimble, and subsonic A-4 first flew in the mid 1950s during the burgeoning era of larger and much more complex supersonic jet fighters then being developed for the U.S. Navy.

The Skyhawk broke the mould, however, by becoming America's first simple, low-cost, lightweight, jet-powered attack aircraft, one that could operate from any size U.S. Navy aircraft carrier then in use.

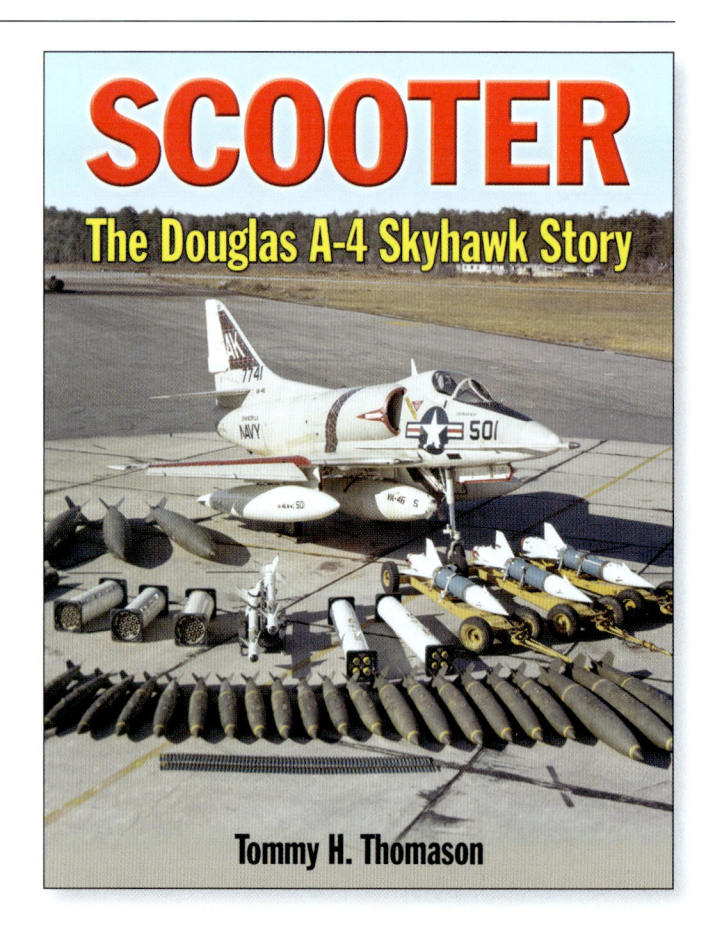

Previous books on the Skyhawk have focused mainly on its colourful combat career, while this book also chronicles, in fascinating detail, the story of the A-4's early years, its subsequent development and its service with at least nine different air arms outside its U.S. service. In their own words, the engineers and pilots who designed and flew the Skyhawk provide exciting new insights into not just the A-4, but also the workings of naval aviation and aircraft carrier operations during the Cold War heyday of the 1950s and 1960s

ISBN: 9 780859 791601

Binding: Hardback

Dimensions: 290mm x 216mm

Pages: 272

Photos/Illustrations: Over 250 b+w and 100 colour photographs

Price £27.95 $44.95

All titles from Crécy Publishing Ltd,
1a Ringway Trading Est, Shadowmoss Rd, Manchester, M22 5LH.
Tel 0161 499 0024
www.crecy.co.uk

Distributed in the USA by Specialty Press,
39966 Grand Ave, North Branch, MN 55056 USA.
Tel (651) 277-1400 / (800), 895-4585
www.specialtypress.com

Other books from Crécy Publishing

US Guided Missiles

The Definitive Reference Guide

Bill Yenne

In *US Guided Missiles* renowned aviation historian Bill Yenne has produced, for the first time, a comprehensive guide to the widely varied United States guided missile systems that have been designated with the "M" prefix.

Beginning with the 1950s MGM-1 Matador – a jet-propelled cruise missile inspired by Germany's wartime V-1 "Flying Bomb" – and the MGM-5 Corporal, evolved from the German V2 ballistic missile, *US Guided Missiles* charts the evolution of Intercontinental Ballistic Missiles (ICBMs) such as the Atlas, Titan, Minuteman and Peacekeeper. The Atlas and Titan later became famous as the basis for the launch vehicles that carried the first American astronauts into space. Meanwhile the RIM-2 and MIM-3 Nike Ajax had their roots in antiaircraft missiles of World War II.

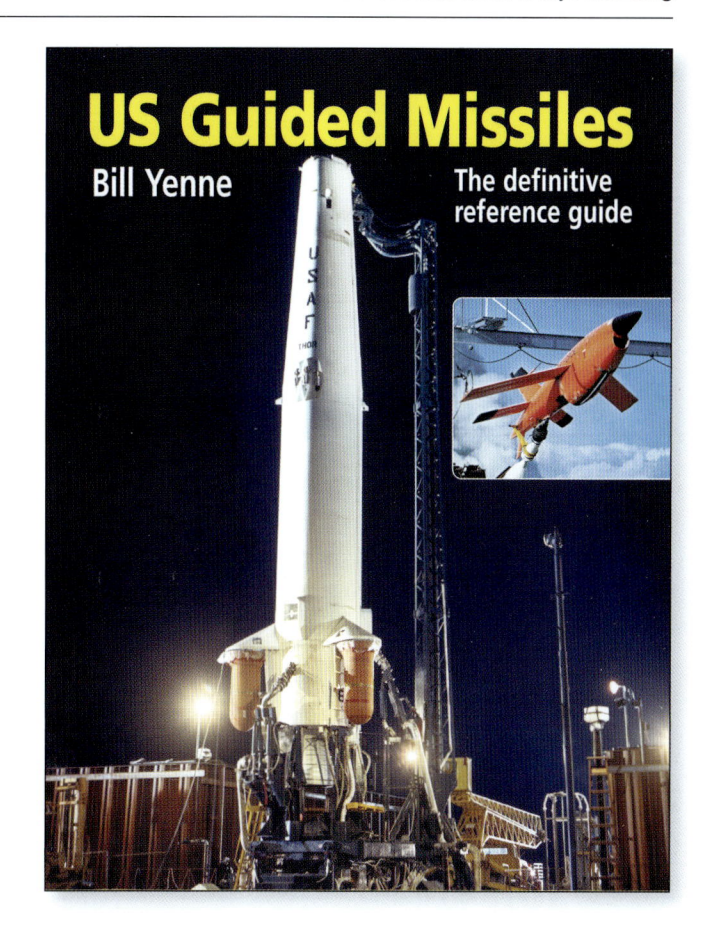

From the earliest Cold War guided missiles, the book progresses through the Submarine Launched Ballistic Missiles (SLBM) including the UGM-73 Poseidon and UGM-96 Trident, to the later cruise missiles such as the BGM-109 Tomahawk. The roster of systems includes the hugely successful air-to-air 'Sidewinder', as well as little-known and obscure missiles, and modern systems in use today including the AIM-120 AMRAAM and RIM-162 Standard Missiles.

Starting with the earliest post-war rockets, through the Cold War to modern weapons, *US Guided Missiles* shows how guided missile systems have changed the face of warfare. Illustrated throughout with rare and previously unseen images, and with extensive appendices, this book is an essential reference for any aviation, aerospace or military historian and enthusiast.

ISBN: 9 780859 791625

Binding: Hardback

Dimensions: 290mm x 216mm

Pages: 256

Photos/Illustrations: Over 200 colour and 45 b+w photographs

Price £22.95 $34.95

All titles from Crécy Publishing Ltd,
1a Ringway Trading Est, Shadowmoss Rd, Manchester, M22 5LH.
Tel 0161 499 0024
www.crecy.co.uk

Distributed in the USA by Specialty Press,
39966 Grand Ave, North Branch, MN 55056 USA.
Tel (651) 277-1400 / (800), 895-4585
www.specialtypress.com

Other books from Crécy Publishing

X-Planes of Europe

Secret Research Aircraft from the Golden Age 1946-1974

Tony Buttler and Jean-Louis Delezenne

In the decades following World War II, aviation designers around the world scrambled to bring new technologies to their aircraft as the jet era ushered in new possibilities and immense challenges. New wing shapes, control systems, engines and construction materials all had to be tested in the air and this required a fleet of so-called 'X-planes' – experimental aircraft designed to test new and untried concepts.

Although much has been written about the legendary American 'X-planes' of this era, far less is known about many of the secret and exotic research aircraft designed and built in Europe especially those in France. Years ahead of their time, these aircraft were world-class in their own right. In Britain the transonic DH 108 Swallow was pushing towards the sound barrier as early as 1946 and the AW52 'all wing' laminarflow aircraft first flew in 1947.

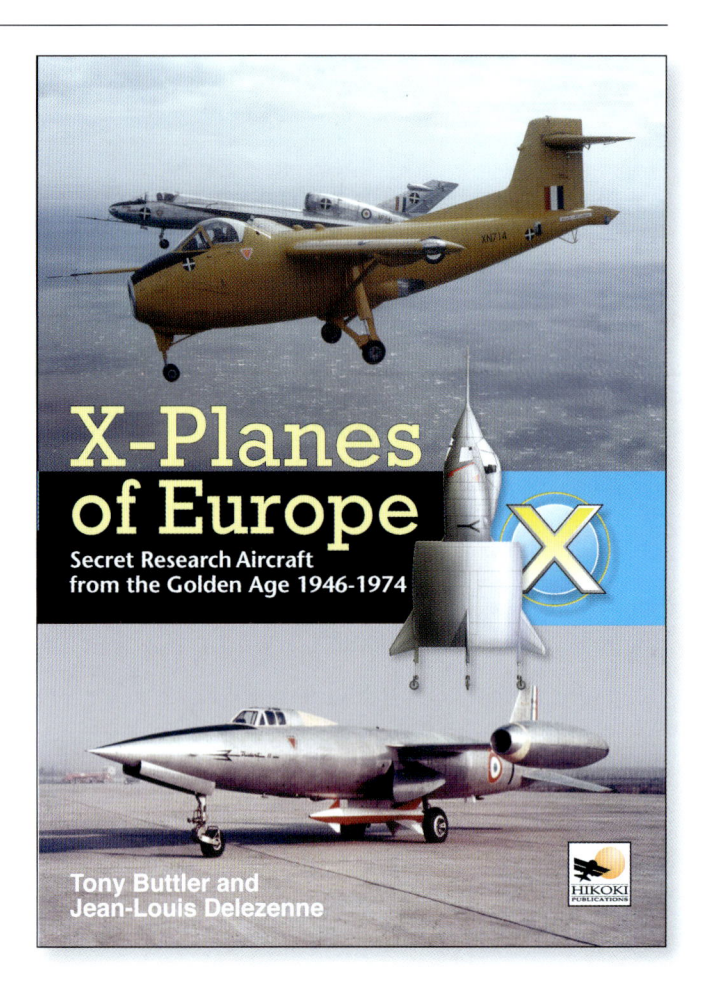

Despite five years of wartime occupation, France flew its own design jet aircraft in 1946, with the N-20 swept flying wing concept being tested by 1951 in Switzerland and the Swedish SAAB 210 delta wing test aircraft first flying in 1952. The Fairey Delta 2 smashed the World Air Speed Record (previously held by a North American F-100C Super Sabre) by an astonishing 37% margin; this aircraft was later pivotal in research for the Concorde supersonic airliner programme. The ungainly Rolls Royce 'Flying Bedstead' of 1953 pioneered the Vertical Take Off and Landing (VTOL) concept used today in the new Lockheed Martin F-35 Lightning II Joint Strike Fighter.

Now, these once highly-classified aircraft are brought together in detail for the first time by the acknowledged experts in this field. The product of years of patient research, much of the material in this ground-breaking book is being made public for the first time. With many unpublished photographs, previously classified drawings and detailed appendices, the stories of these remarkable aircraft combine to produce an in-depth record that gives these rare and exotic flying machines their proper place in aviation and military history.

ISBN: 9 781902 109213

Binding: Hardback

Dimensions: 297mm x 210mm

Pages: 304

Photos/Illustrations: Over 250 photographs and illustrations

Price £34.95 $56.95

All titles from Crécy Publishing Ltd,
1a Ringway Trading Est, Shadowmoss Rd, Manchester, M22 5LH.
Tel 0161 499 0024
www.crecy.co.uk

Distributed in the USA by Specialty Press,
39966 Grand Ave, North Branch, MN 55056 USA.
Tel (651) 277-1400 / (800), 895-4585
www.specialtypress.com